AF664055

# AGRO-ENTREPRENEURSHIP

## About the Author

**Dr. Shakti Ranjan Panigrahy** has been working as Assistant Professor in Department of Agribusiness Economics and Policy since December 2013. He holds a master degree in Agribusiness Management from Chaudhury Charan Singh Haryana Agricultural University, Hisar. He has cleared UGC (NET) in Management (2012), ICAR (NET) in Agribusiness Management (2013) and SRF with Fellowship in Agribusiness Management (2014). Dr Panigrahy had served in Fishery & Animal Resource Department, Government of Odisha for almost six years as Veterinary Assistant Surgeon. He has five research paper, eight popular articles in his subject area of Agribusiness. His maiden book, "Objective Agribusiness Management" is also circulating in the market. This book will be his second venture in this segment for the academician, researchers and upcoming entrepreneurs.

**Baljeet Singh** holds a Masters in Agri (Horticulture) from BHU and an MBA in Agribusiness Management from MANAGE, Hyderabad. He has more than eight years of experience in both public and private sector organizations of repute like Anand Agricultural University, SD Agricultural University, Dantiwada, Monsanto and IRMA. Baljeet's interest areas include sales, marketing of agricultural inputs, agribusiness incubation and technology transfer. Currently he is pursuing a Doctorate from GH Patel Post Graduate Institute of Management Studies, Sardar Patel University, V V Nagar, Gujarat.

# AGRO-ENTREPRENEURSHIP

**Shakti Ranjan Panigrahy**
Assistant Professor
International Agribusiness Management Institute
Anand Agricultural University
Anand, Gujarat-388110

**Baljeet Singh**
Doctoral Scholar
GH Patel Post Graduate Institute of Management
Sardar Patel University
V V Nagar, Gujarat

SCIENTIFIC PUBLISHERS

Published by
**SCIENTIFIC PUBLISHERS (INDIA)**

Jodhpur –

5 A, New Pali Road
P.O. Box 91
Jodhpur - 342 001 INDIA

Delhi –

4806/24, Ansari Road
Daryaganj
New Delhi - 110 002 INDIA

© 2017, Scientific Publishers (India), Jodhpur

All rights reserved. No part of this publication or the information contained herein may be reproduced, adapted, abridged, translated, stored in a retrieval system, computer system, photographic or other systems or transmitted in any form or by any means, electronic, mechanical, by photocopying, recording or otherwise, without written prior permission from the author.

**Disclaimer:** Whereas every effort has been made to avoid errors and omissions, this publication is being sold on the understanding that neither the editors (or authors) nor the publishers nor the printers would be liable in any manner to any person either for an error or for an omission in this publication, or for any action to be taken on the basis of this work. Any inadvertent discrepancy noted may be brought to the attention of the publisher, for rectifying it in future editions, if published.

**Trademark Notice:** Product or corporate names may be trademarks or registered trademarks, and are used only for identification and explanation without intent to infringe.

ISBN: 9789386102386

ISBN:

9 789386 102386

Visit the Scientific Publishers (India) website at
http://www.scientificpub.com

Printed in India

## *Dedication*

*This book is dedicated to*

*My Parents (Both), My Wife (Arpita) and*

*My Beautiful Daughter (Gidi)*

**SRP**

*This book is Dedicated to*

*My wife (Nibha) and daughter (Rubal)*

**BS**

# Preface

Agriculture is backbone of India. Today, India has become a net surplus producer and exporter of food grains and is next only to China in fruits and vegetables production. However, number of people taking up agriculture is reducing day by day. Agri processing sector is also not very well developed unlike in countries like Japan and Malaysia where a large chunk of agri-produce is processed. It also means that investments in agri enterprises are low. Reasons could be many ranging from complexities of starting a new business, lack of capital, lack of technical knowhow to risks associated with new agribusinesses etc.

Every year, more than 10,000 agriculture graduates and 5500 post graduates pass out of Agricultural Universities. Most of them take up higher studies or end up working for someone else (private companies, government organizations etc). One mandate on which our great temples of learning are losing out is preparing enough number of risk takers who can start their own agribusiness. These should aim at producing more of job providers rather than job seekers. This will ease the burden on many other students.

Almost all State Agricultural Universities have an MBA Agribusiness College. Placement is a tough time for students in these colleges. Government cannot provide jobs to all, neither can private sector absorb everyone. Competition among themselves is one reason for it, there are older standalone institutes like MANAGE, NIAM, XIMB, SIIB, CMA (IIMA), IIML etc who have been doing pretty well since decades and most of good companies end up recruiting from these institutes. In such a scenario, the best option is to go for one's own agri enterprise. However, this road is not laden with red carpet. The problem is aggravated by the fact that agriculture is dependent on nature. On the other hand food is a basic necessity and people will any way spend on food. Other way to look at this is- every challenge presents a business opportunity. Nevertheless, entrepreneurship is charming as well. There is no boss to an entrepreneur and he is the master of his time.

The present book is a product of many brains, most of whom have practically been facilitators of agri entrepreneurship in one way or the other. The book will be helpful to aspiring and budding entrepreneurs who want to make it big in agri-preneurship. The book mainly focuses on how the challenges of startup agribusinesses can be overcome in a cost and time effective manner. The book will prove to be a facilitating campanion for everyone in startup agri-preneurship fraternity.

The book is divided into five major units in which total twelve chapters are included. The first chapter is an introduction and sets into excitement, young agri graduates and post graduates to try their hand in entrepreneurship. The second to ninth chapters are on entrepreneurship opportunities in Agriculture sector and delve upon Agri Biotech, Food Processing, Protected Cultivation, Floriculture and Landscaping, Supply Chain Management, Agri Retailing, Livestock and allied sector. Tenth chapter is on project formulation and will guide the readers in a step by step process to new agribusiness project. Eleventh chapter is on support system on Agri Entrepreneurship and focuses on both Government and private sector support to agri entrepreneurship in the country and last chapter brings together some existing innovative start up models in agri entrepreneurship from across the country. We hope that the book will be useful to all interested in agri entrepreneurship.

The authors are thankful Ekta Jajoo of Onganic, Praful Patel of Agribiz Corporation, Ishira Mehta of Crop Connect and Titu Yoko of Natur Oma who shared their innovative models of agri entrepreneurship and took time out of their busy schedules to answer authors' queries. We are also thankful to various other agro enterprises, agribusiness companies, agri business incubators both in the private and public sector who extended useful information to bring this book in present form. We are thankful to Scientific Publishers (India), Jodhpur for timely printing of the book.

We are very much thankful to International Agribusiness Management Institute of Anand Agricultural University for providing us an academic environment without which this book could not come in its full structure.

This book is just the beginning of a literary revolution in agri entrepreneurship and is in no way complete. Useful suggestions and constructive comments are always welcome for improvement of the book. We would be more than happy to have the errors pointed out so as to be corrected in the next edition.

**Shakti Ranjan Panigrahy**
**Baljeet Singh**

# Contents

*Chapter - 1*

# AGRO-ENTREPRENEURSHIP - AN INTRODUCTION

**Shakti Ranjan Panigrahy**

*Assistant Professor, IABMI, Anand Agricultural University, Anand, Gujarat*

## DEFINITION OF ENTREPRENEURSHIP

Before discussing much about entrepreneurship, we should know what the word an enterprise means. A project that requires boldness and energy due to its complex nature is called **enterprise**. An indivisual who takes innovative and profitable steps for his venture and manages risk associated with its step is called an **entrepreneur** and this whole activity is called the **entrepreneurship.** The term, **entrepreneurship** came from the French word *entreprende* which signifies to do something without any link to economic success. However, the meaning of entreprende has no significance in the competitive world. But to survive in the business one has to think out of the box. Otherwise he may perish in the red ocean.

Drucker in 1985 perceived entrepreneurship as the creation of a new organization, regardless of its ability to sustain itself, let alone make a profit? That's why the vision of an indivisual turns out to be a venture i.e. the vision full of risk and innovation. Innovation is a complete and a continuous procedure; for that sake an entrepreneur should alway innovating products and services on frequent intervals to make a niche in competitive market that also helps in satisfying the demand of the customers.

World is not increasing fast only in terms of globalization and urbanization, the concept of nuclear family is also in trend. That's how an entrepreneur understood the importance of increasing growth of nuclear family. The entrepreneurs started to think and invent miniaturization in agricultural products by processing of garlic and ginger paste as HOMMADE (Dabur) or a product like Maggi (Nestle) that is being prepared within two minutes. These innovations proved to be a boon for nuclear family that gives a different shape to the concept of entrepreneurship.

## ANATOMY OF ENTREPRENEUR

The study of complete structure of a human body is called **anatomy**. Entrepreneur's anatomy consists of three basic elements i.e. knowledge, passion and fire. When the data is processed, it is the **information** that has some socioeconomic significance and is called **knowledge** (Fig. 1.1). To study the impact of knowledge from the agro entrepreneurial perspectives; we should understand its significance both in manufacturing and service sector.

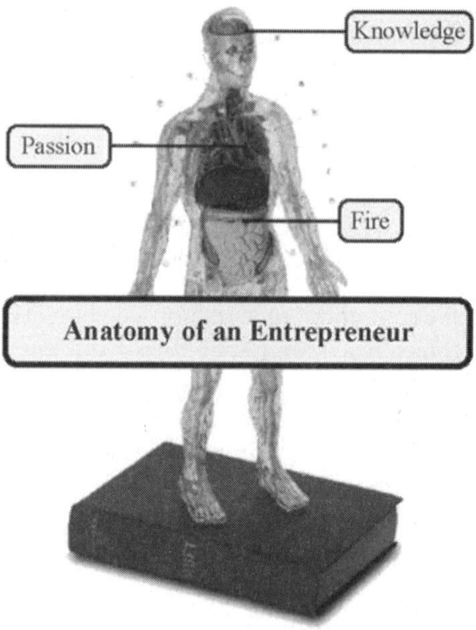


**Figure 1.1:** The anatomy of entrepreneur

### Caselet

***Infosys, IT Service Industries***

Infosys established its knowledge management department with a philosophy, *Learn Once, Use Everywhere*. In its K–shop, all the employees interact through web interface (Sparsh) Infosys motivates its employees by providing knowledge currency unit for their innovative contribution in the knowledge asset of the institution. Infosys circulates knowledge directory as people knowledge map in which any employee can find the address and location of the experts on quick basis. Infosys also circulates knowledge currency unit scheme for rating the quality of assets in the repository. All these activities make Infosys as a first Indian IT firm in the list of NASDAQ.

***Dr Reddy's Laboratories***

Dr Reddy's, a pharma manufacturing company, has also established its knowledge management architecture. They have also developed an idea box as a valuable database store. Dr Reddy's Laboratories has invested in an e-campus, where one of the modules is focused on safety, health and environment (SHE) issues and is targeted at the shop floor supervisors. There is a knowledge management grant for Rs 1500 available for every employee for purchasing books or other materials every year. It was felt that knowledge management reinforces employees in a positive way as a result its impact is observed in productivity and competitive advantages.

The sole objective of demonstrating these inspirational examples under caselets is if we can generate name and fame by encashing our knowledge base in manufacturing and service sector; then why it will not be possible in agriculture. It is possible, only when we transform our agriculture to agri-business by putting flavor of innovation in the market. Exploring right market for the right product is the key of today's business. From production center to consumption, we need empowered human resources for getting competitive advantage and for that sake inculcating knowledge and information in the avenue of entrepreneur is necessary. It is well known that more than 50% of labour is dependent on agriculture and if we are able to transform them in to agroentrepreneur by continuous knowledge dissemination then agriculture will become the largest industry in India.

The second important element is **passion**. Passion is a term applied to a very strong intuitive feeling for a particular thing by an individual. Similarly fuel in the fire of ambition, determination, dedication make an entrepreneur more skeptical and make him more capable to tackle the different uncanny situations. Passion for being a leader, independent decision maker, owner of your own entity motivates many individuals to start their own enterprise and to be a fearless entrepreneur.

## Caselet

***Grows micro-greens in soil-less medium***

This is the business marathon of Mr N Vidyadharan and his micro-green firm, **Aditya Organic**. From social worker to real estate consultant; and from travel agent to organic fertilizer producer, he travelled many paths but he found his own destiny by setting up his firm in the urban area of Chennai. When his organic produce (liquid organic fertilizer) had not been accepted by any one after several efforts, he thought to try this input himself in his firm. Inspite of various uses of micro-greens for garnishing salad and improving nutritious aspect of food items along with its application as fertilizer, the best effort of Mr Vidyadharan could only sell eighteen varieties out of thirty trial varieties initially in the market. Currently he sells micro-greens in fifty boxes, but target is for two hundred boxes. So far the capital investment, he has spent almost Rs. 35,000/- and is getting Rs.18,000/ to Rs. 20,000/ per month. If he achieves 200 boxes target then income may exceed Rs.70,000/ per month. Although lots of difficulties are there in the marketing of organic liquid fertilizer still a persistent innovation can create a niche area for a product.

The most interesting question comes whether an agricultural entrepreneur is different from non-agricultural entrepreneur? The answer is yes because in agriculture, entrepreneurship is surrounded with claws of local product, vagaries of climate, post-harvest variability, infrastructural accessibility and market uncertainty. Women agro-entrepreneurs involvement are more than men may be due to their fitting in ability, risk tackling power and multitasking aptitude proves better in comparison to their counterparts. Another peculiarity of agro-entrepreneurship is the family firm orientation. Unlike non-agricultural entrepreneur whose major objective is to garner profit and growth; family firm orientation is more towards the survival, family heritage, autonomy, rural life styles and passing a healthy farm to the next generation.

## AGRO ENTREPRENEURSHIP-ITS ROLE IN PRESENT SCENARIO

India is an agricultural based country where more than 80% of farmers are small and marginal. It means their land holding capacity is less than 2 hectare and it is observed; in India, per capita land holding is just about 1.16 hectare. This is the reason; although we are number one in production of many commodities yet we are far behind in our productivity status than our counterparts. The agricultural production deteriorates due to overwhelming population, land degradation, ever changing dietary status; and uncertainty in climatic conditions that put more pressure on our productivity. In this scenario, we can sustain only when we go for production by masses instead of mass production.

It has been observed that demand for food grain production is inelastic whereas demand for fruits, vegetables and for milk, milk products is showing elasticity. In India, more than 60% population are young, ambitious, risk takers and independent decision makers which separate themselves from the so called **traditional agriculture** where self-sustenance was a major issue to commercialized agriculture where many activities are organized and profit making is the sole objective. The turn over period in agriculture is more than the industry; as a result entrepreneurs move towards other sectors such as dairy, and fishery. On the other hand the establishment of agro-industry in the rural areas puts check on migration and accelerates employment. As a result social harmony will be maintained as the indicator of growth.

We can improve our agricultural productivity either through innovation in products or in processes or through minimization of post-harvest losses (Fig. 1.2). Apart from innovation and strategical minimization of losses the technological up-gradation and improvement of infrastructure is also necessary. Since the time of

Independence, government is acting for governance and private sector for productivity and Indians were succumbed with the paucity of funds, malnutrition, low level of employment, inequalities of income and many more challenges. To safeguard the interest of the population, public sector came forward as there were few players at that time that had both managerial and scientific skills along with risk aversion capacity with a long gestational investment. In these instances, public sector took the charge of significant development in agricultural and industrial base, diversify the public economy and overcome the economic and social backwardness.

Public sector controls the activities of the private players through Monopolies and Restrictive Trade Practices (MRTP Act 1969), higher side corporation and income tax. But due to acute shortage of fund and severe fiscal deficit, public sector went for liberalization, privatization and globalization in 1991 and handed over the charges of development to the private players. After that, private sector approached the customers with a tenet of quality, price and service (QPS). High quality, low price and high service were the tool to earn the customer loyalty. At the time of 2001-2002; to attract more customers, private sector approached merger and acquisition, new geographical penetration, diversification like strategies but again fiscal deficit came in the period of 2007-2008. Then both the public and private sector thought that the current situation like fiscal deficit, unemployment, inflation etc only could be under control, if the policy moved more towards innovation, integration and inclusiveness. That's why government has always tried to bring the transparency in the policy making to garner more transparency and accountability among the investors and it was at its peak during these years. If the faith is built on, then automatically it will bring investments. Another important question such as who will invest in the market. Who has enough strength to feel the thrill of venture that is full of risk and uncertainties? As market is more tensed and fragmented than before due to severe competition, only persistent innovation is necessary for survival, growth and profit.

Innovation has different meanings for different people. At same platform invention of aeroplane is an innovation and invention of parachute is also an innovation.

In this section, we discuss the agricultural innovation. From very beginning, we has emphasized on productivity under the clutches of climate and competition (Fig. 1.2). Magnetic agriculture is an innovation in the direction of agricultural production where use of saline water in agricultural production can decreases the hardness of tubewell water and makes it useful for Rabi crop production. Nanotechnology in agriculture not only helps in slow release of fertilizer but also reflects the quality of food products by changing color of the package. Now we movetowards organic agriculture, an innovation of our forefathers; as they used organic input materials for their agricultural production. Though it is the need of this century to produce organic food and maintain sustainability, yet it is a metaphor of grass root innovation. Green milk is a recent innovative product invented by CFTRI where they use moringa, kheera, tulsi, and mushroom like materials and targeted to both vegetarian and vegans. Now we market the products which are environmentally preferable. In previous London Olympic, Ice cream in the coconut cup was a big favorite among the customers. They accepted it as an innovation.

**Figure 1.2:** Innovations in agriculture

Before going to the innovation in agriculture, we should know our agricultural potentiality. In India, agriculture contributes almost 13.7% to its GDP where as its annual growth is all about 3.3%. Almost 52.1% of Indian people either directly or indirectly depend on agriculture (Fig. 1.3). If this is the condition; then we cannot eradicate the poverty cycle and disguise unemployment in present circumstances.

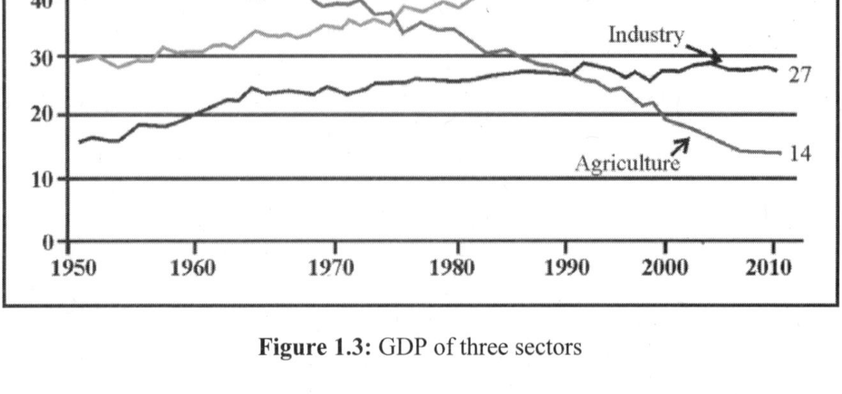


**Figure 1.3:** GDP of three sectors

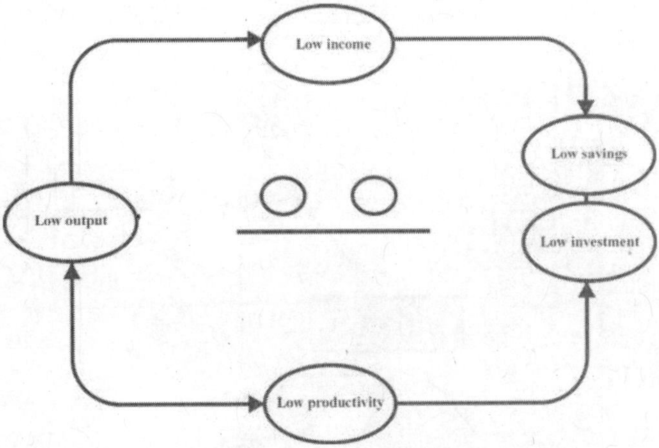


**Figure 1.4:** Poverty cycle

Figure 1.3 signals decreasing trend in GDP contribution of our agriculture and Figure 1.4 signals poverty cycle due to less productivity; we cannot improve our income generation ability which affects our saving, investment, capital generation. It is a continuous cycle and our farmer is always remains poor one due to this. That's why, new innovation and new technology in our agricultural production system is necessary for the benefit of growth and development.

## Caselet

In initial days, farmers are depended on stakeholders for agricultural input in the form of seed, fertilizers and on government for consultancy as well. Similarly, during marketing of its product, they also depend on kachha and pakka adat. This cobweb system reduces the value of the product for the producer. Understanding this problem, ITC established a model named e-choupal in 2000. They conceived an idea from village square- a kind of place we call choupal in North India. Here, people gather to smoke hookah, watch TV, exchange gossip, laugh a while, and talk about work and the weather, before they finally head home. However, the moment, this place is enabled with a PC, provided access to the internet, called an e-choupal; a business model around it gets built in a uniquely Indian way (Fig. 1.5). Here ITC started selecting lead farmers (Sanchalak) to head each Choupal. His work is to provide information on weather and market price, scientific farm practices, risk management, sale of farm inputs and purchase of farm produce from the farmer's doorstep. A new role was envisaged to the commission agent of Mandi (Samyojak or Coordinator). They were asked to not only use their ties in the villages to nominate sanchalaks, but also made responsible for relevant mandi documentation. Samyojak would aggregate the grain and bring to ITC to get 1% commission. This way value of the product was increased for the farmer and information is channelized from authenticated sources and the farmer becomes empowered by right information.

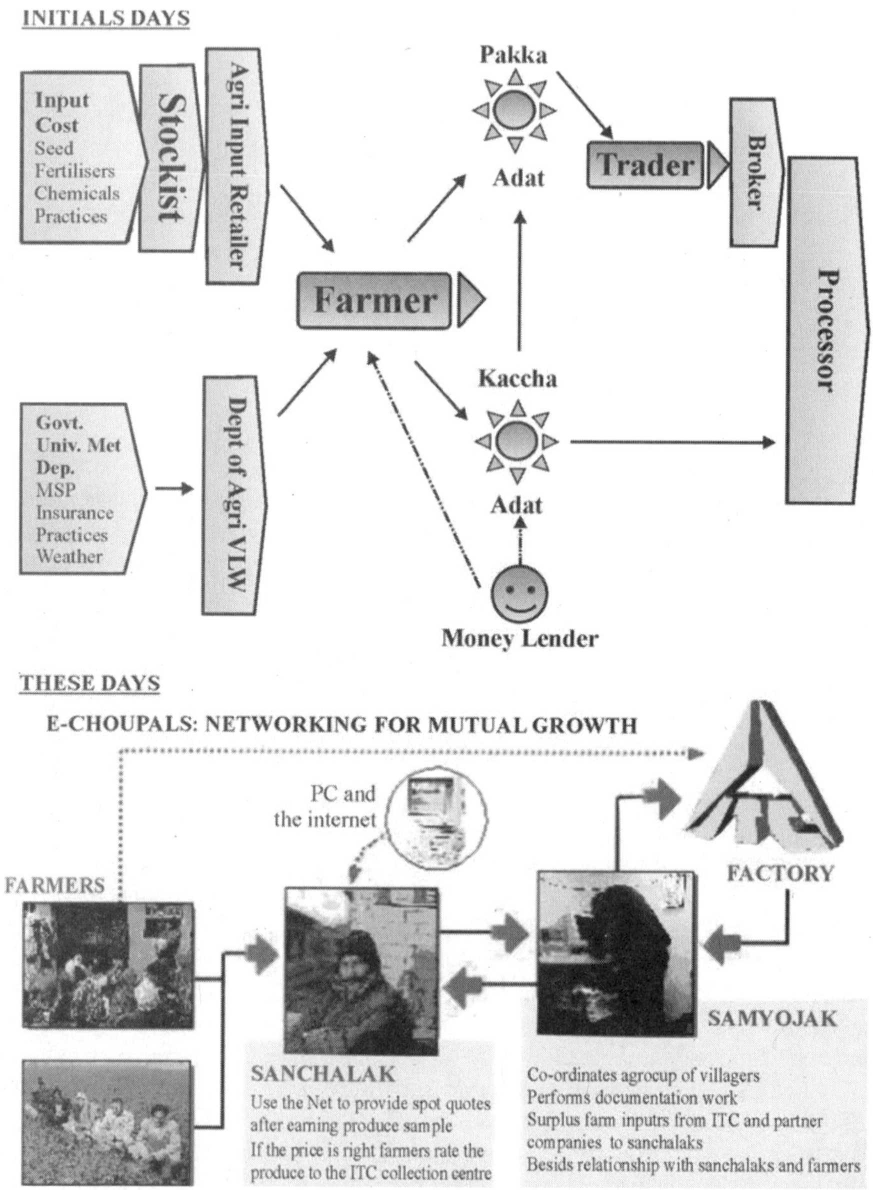


**Figure 1.5: E-choupal networking**

It is important to discuss how integration is necessary for sustained growth in agriculture. Integration of resources from man to material is required to harness economies of scale. Self help group (SHG), cooperatives, producers companies are some of the integrated models in agricultural sector. Through these institutions, the vulnerable sections of the society find their way towards development. Everyone likes to listen to the story of rags to riches. But few of them can make it a path breaking story. One of them is the Shri Mahila Griha Udyog Lijjat Papad as discussed below.

**Caselet**

***Shri Mahila Griha Udyog Lijjat Papad***

This is the case study of sheer determination, empowerment, and organized movement of a group (Cooperative). It is a cooperative that had come into action in the year, 1959 with a seed capital of Rs. 80 and membership of only seven; but now it has crossed almost Rs. 300 Crore in its annual sales turnover. This is popularly known as lizzat, an Indian woman cooperative. In every household use of papad is common as a pre-food snack; India's most popular crispy bread. This cooperative is based on Gandhian simplicity where work is distributed among the workers from collection of dough to production of papad; and from testing and packaging of papad to distribution and collection of vanai (rolling charge). Here membership only can be provided to those women who have respect for quality and for the institution. For rolling of papad, a clean house is necessary. Those not having this facility are engaged in collection of dough and distribution of papad. Decentralization of activities in all geographical areas of India makes the product non uniform as different regions have different varieties of urad wheat. That is the cause lizzat papad has no USP. But a uniform policy and integrated approach to bring the women in to a group philosophy; make them expand from local to global. Innovative approach like circulating lizzat patrika among the members at a subsidized rate, providing Chhaganbapa Smruti scholarships to the daughter of member sister makes the cooperative more trustworthy, empowered and successful as transparency restores faith and strengthens business.

Similarly inclusiveness is one factor which states that each pie of developmental success should reach every nook and corner of the society. Business stands on two policies at the juncture of severe competition. Either you try to increase the usability or you try to motivate them to use more at a time. For a businessman to look at the grass root level is a must awaiting step as the current market is saturated in the urban areas. To maintain the overall balance between local and urban, government is also establishing infrastructural facility like road, electricity, irrigation, storage, warehousing and many more at the grass root level to attract investment and innovation. Then each activity sparks into a new business and develops a new entrepreneur. Financial inclusion is the mantra of these days as it is necessary to connect everyone with the mainstream. Hopefully, it establishes transparency, predictability and accountability- the three pillars of democracy.

## Caselets

### Agro-entrepreneurship showing inclusiveness

- **mKrishi** is a software developed by Tata Consultancy Service in which farmer contacts with the service provider companies instantly and takes the benefit of good quality agricultural inputs, finds advice on farming practices, and gets information on market prices, weather and other information that can help in planning and boosting yields. Better information reduces the uncertainties in agriculture and enhances overall shareholder wealth maximization. TCS not only creates its brand value by presenting mobile based services for farmers as well as different national and international agencies but also, it generates a platform for different micro-entrepreneurs at the grassroots level.
- **Reuters market light** (RML) is a subscription based mobile phone messaging service which provides highly customized and localized agricultural information to Indian farmers. It provides messages regarding the local weather forecast, local market prices, relevant news for the two types of crops chosen, information about harvesting, proper fertilizer usage, irrigation methods, seed and soil combinations and other advisory tips to improve productivity.
- **Nokia life tools** is an SMS based service which provides interface to reuters market light, which delivers information on crop prices and has enabled farmers to find out how much they can expect to sell their crops for strengthening their hand when negotiating with local agents.
- **Ikisan** developed by Nagarjuna Group is an agricultural portal, which provides online, detailed content on crops, crop management techniques, fertilizers & pesticides, weather forecast and a host of other agriculture related information.
- **mKisan** is a mobile based agriculture advisory service implemented in India by Handygo Technologies, CABI, South India and Digital Green, and NGO. It consists of agriculture advisory from experts, agro bulletins, market information on crop prices, weather forecast and a farmer helpline.
- **Avaaj Otalo**, developed as the collaboration between UC Berkeley School of Information, Stanford HCI Group, IBM India Research Laboratory and Development Support Center (DSC), an NGO in Gujarat, is a service for farmers to access relevant and timely agricultural information over the phone. By dialing a phone number and navigating through simple audio prompts, farmers can record, browse, and respond to agricultural questions and answers.

In the era of innovation, integration and inclusiveness, the role of agro-entrepreneur is a demandable component that helps to improve the productivity by reducing wastage, channelize information, establishing infrastructure, enhancing employment, maintaining socioeconomic balance by accelerating production and productivity particularly in agricultural sector.

## TYPES OF ENTREPRENEUR

### Innovative Entrepreneur

An innovative entrepreneur is an individual who adds value to a particular niche or market sector by providing innovative products or services that are in demand. Innovative entrepreneur can be defined as the effective combination of labour, capital and property utilized to create innovative products in keeping with the changing demands of the market.

This question arises who will innovation because innovation needs cost and these days market is more flux and fragile. In developing and underdeveloped countries though consumer surplus is less or not sufficient to accept an innovative product or service, so enterprises either reduce the price of the product or waive off from that market by not fulfilling the overhead cost. Besides these, to generate innovation in product, a substantial degree of knowledge, skill and aptitude is necessary. When policy of the government is not amicable then also innovation cannot find its proper path. That's why, innovative enterprises are emerging more from the developed countries as business environment provides the entrepreneur, a risk averse platform.

### Imitative Entrepreneur

Imitative entrepreneurs are characterized by readiness to adopt successful innovation inaugurated by already successful innovating entrepreneurs. Imitative entrepreneurs do not imitate the changes themselves, they imitate the techniques and technologies innovated by others.

According to Michael Porter, development of enterprise is either through factor-driven or through efficiency-driven or through innovation-driven. In the **factor-driven** stage of development, as large numbers of unemployed people are there in the informal sector, it pushes small scale industries to grow and in **efficiency-stage** of development, use of production facility efficiently leads to growth. These two areas of growth attract entrepreneurs to invest in developing countries by imitating developmental prototype of developed counties. In developed countries innovatively driven enterprises are found due to balanced use of resources. When innovation has been carried out, different trade related intellectual property rights come into picture to safeguard the interest of the inventor or entrepreneur through patent, copyright, trademark like procedures. If someone wishes to use this innovation he will have to go through different licensing procedures.

### Fabian Entrepreneur

A Fabian entrepreneur is one who is very cautious in taking the decisions, very skeptical, and takes calculative steps and basically showing somewhat lethargies for accepting a new innovation. Some examples include rich farmers and young people seeking employment. In some instances it has been observed that the first generation of a business family is Fabian entrepreneur. They follow the footsteps of their successors. They imitate only when they are very clear that failure to do so would result in a loss of the relative position in the enterprise.

## Drone Entrepreneur

Drone entrepreneurs suffer losses, as they refuse to make any modifications in the existing production methods. These entrepreneurs are conservative or orthodox in their outlook. They never like to get rid of their traditional business and traditional machineries or systems of the business. They always feel comfortable with their old-fashioned technology of production even though the environment as well as the society have undergone considerable changes. They are laggards as they continue to operate in their traditional way and resist changes.

## Copreneur

Copreneurs are the enterprising couples who work in and share ownership of, commitment to, and responsibility for their business. By couple it has been stated that this could be a relationship that is married or unmarried but they jointly share all aspects of the business.

### Caselet

***Bamboo House India***

It is a case study of green livelihood and sustained entrepreneurship between a copreneur and tribal artisans who depend on bamboo sector. Prasant Lingam, The Husband and Aruna Kappagantula, the wife while searching for an eco-friendly furniture, found great difficulties to get it in India as in that time (2008) bamboo furniture was being imported from China, Indonesia, Malaysia etc. After rigorous searching on the internet, they found more than 50% of North-east Indian population depends on bamboo for their livelihood. After doing proper feasibility study, they got to know that Tripura is the only state where the best bamboo species like *Thyrsostachys oliveri* or Kana kais are found. The peculiar characteristic of this type of bamboo is that its height is 60-80 feet and diameter is 15-55 mm, which is achieved in four years. The other varieties like *Aleverii* and *Balcoa* are found in Tripura as well and stricter is found in Nagaland and Mizoram. The couple also found that the inhabitants of village of Katlamara of Tripura are highly skilled in furniture making from bamboo. Before starting the business, they conducted an exhibition under the assistance from Andhra Pradesh technology development and promotion centre and found a great response from the public. The couple made a linkage between the demand of Andhra Pradesh and supply area of Tripura, Mizoram and Nagaland and started a business venture called as Bamboo House India. Though lots of challenges were there from infrastructure (road, telephone connectivity) to language barrier (workers are bangla speaking), still the couple handled the business cautiously to make it a successful business model.

***Harvest Gold: A bread based enterprise***

This is a case study of a copreneur, Mr. Adil Hasaan (Husband) and Taab Siddique (Wife) who came to India for settlement but did not get good quality bread for consumption as they were comfortable with the European and American based food habits. They thought to provide bread at a lower cost and improved quality as they understood that entry to bread market is quiet easy. That's why; they thought to march on the business with a philosophy of One's plant, one city and one product. They generated their core competency for their product through improved quality (as their plant is near Bhiwadi where they got a plenty of raw material, wheat of improved quality), positive perception through humorous advertising could accelerate availability through their own truck. Besides that cellophane packaging and humorous print advertisement in Delhi Times on Fridays makes their product accessible to the consumer. In this way, both these copreneurs made their business successful slowly and steadily.

## Intrapreneur

Gifford Pinchot (1985), one of the originators of the term intrapreneurship had observed that intrapreneurs are people who dream beyond their mundane domain of something unusual. Intrapreneur is also called an **inside entrepreneur**, or an **entrepreneur** within a large firm, who uses entrepreneurial skills without incurring the risks associated with those activities. Intrapreneurs are usually employees within a company who are assigned a special idea or project, and are instructed to develop the project like an entrepreneur. They usually have the resources and capabilities of the firm at their disposal. Their main job is to turn that special idea or project into a profitable venture for the company. Intrapreneurship is also called as **corporate entrepreneurship**.

The environment of intrapreneurship mainly reflects from the overall culture of the organization. The organization should be more flexible in its working pattern, should provide immense strength to the employees taking innovative steps, should invent and erase unproductive cogs in the developmental wheels, and should delegate authority in accordance with the responsibility. This type of working cultures generate intrapreneurship in the organization as in today's era of globalization past entrepreneurial set up always faces challenges to survive and today's intrapreneural outlook takes the organization from the scratches. Freedom to fail and make a little and sell a little is the basic tenet of all the intrapreneurs. In true sense, resource constraint and time constraint create intrapreneurship in the organization and they always search prosperity at the advent of adversity.

After the advent of globalization, it was necessary to bring the momentum in environment of competition, collaboration and communication. That's why organization put investment on different projects for bridging the gap between the actual and target achievement. The way to create innovation for undertaking project for survivability, profit and growth gave a new thrust to the organizations in the form of intrapreneurship. In this dynamic world in which we live it is intrapreneur who is more important than entrepreneur. No doubt an entrepreneur creates an organization but it is intrapreneur who drives it into glory. If it is not true then why many organizations became sick after globalization and closed down their shops?

## QUALITIES / TRAITS OF AN ENTREPRENEUR

An entrepreneur to be successful needs to have certain traits which are briefed below:

1. **Positive attitude and being optimistic:** A successful entrepreneur keeps a positive attitude towards life and creates an aura around him/ her to encourage and motivate others as well. However, he does make a sensitivity analyses for his projects.
2. **Stress bearer and hard worker:** Path of entrepreneurship is not laden with roses. A successful entrepreneur has to work under pressure, under strict deadlines to take advantage of opportunities. He has to be a hard worker. Striking a fine work life balance may at times be difficult for him
3. **Systematic planner**: A successful entrepreneur has great planning skills. It is utmost important in changing times as economical, political, social and technological environment is changing so fast.
4. **Independence:** A successful entrepreneur is independent. He is able to take his own decisions with the available information.
5. **Alertness to opportunities:** A successful entrepreneur is always watchful for new and emerging business opportunities and takes advantage of these. In fact, they should see every challenge as an opportunity.
6. **Calculated risk taking:** A good entrepreneur takes calculated risk. He uses all available information to make the best decision which maximizes his chances of winning.
7. **Winner:** A successful entrepreneur has a die-hard winning attitude. He takes best suitable actions to convert his ideas into achievements.
8. **Leader and follower:** A good entrepreneur is both a leader and follower depending on the situation.
9. **Confident:** A successful entrepreneur is bubbling with confidence and acts as a source of inspiration for others as well.
10. **Ownership:** A successful entrepreneur has a sense of ownership. He is accountable and responsible for his actions.

## SIGNIFICANCE OF ENTREPRENEURSHIP IN ECONOMIC DEVELOPMENT

Before we discuss anything about the economic development, we should understand the basic picture of different nations. In under developed countries, we are confronted with the scenario of unemployment, high labour land ratio, disparity between the rich and the poor, low savings, low investment, dualism in techno-social outlook, poor entrepreneurship and excessive dependency in agriculture. As a result, per capita income in under developed economy is low and the basic resources like manpower and natural elements are underutilized and unutilized. But in developed countries like USA, Australia etc the opposite conditions are seen. The well-established technological development and entrepreneurial outlook improves the productivity in various sectors and employment generation from manufacturing and service sector make their national economy stronger than the under developed countries. It means when the under developed and developing countries like India face difficulties to eradicate the basic problems in their developmental scenario, in that instance developing countries carve the growth cycle through their innovative and entrepreneurial approaches. To substantiate the developmental process in the right track; we should give more emphasis on grass root innovation.

It has been observed that in developed countries recession comes when their capital resources are not utilized (means they have no demand for goods or services); but in under developed and developing countries recession comes, when they do not generate enough capital resources for sustained development. This capital generation is only possible when we give ample opportunity to the masses instead of classes. These masses are the entrepreneurs, who not only produce capital resources but also channelize even distribution of resources. As a result decentralization of economic power, capabilities, responsibility, and accountability can be established.

## Entrepreneurship is a Labour Intensive Activity and it creates Jobs

India needs more than 60 million additional jobs by 2020–twice the job addition between 2005 and 2010 to maintain the current ratio of employed people to total population of 39 %. With millions of people employed today either retiring or losing their jobs, new job hiring will have to exceed 60 million by 2020. In a report of ALjAZEERA America on 18 May 2014, it is reported that in India almost 12 million are estimated to enter the job market each year. In these circumstances it is not possible to generate employment solely by government or by private sector. Entrepreneurship is the tool that creates jobs because these days government policy like single window system, quick clearance, and faster decision, investments in infrastructure and about Rs 10000 crore on venture capital scheme attracts many entrepreneurs to invest. But how actually entrepreneurship is able to create jobs and utilize labor in its activities, is clarified from the following case studies.

### Caselet

***Dabbabala- An example of grass root innovation***

A dabbabala, an indivisual who collects hot food in lunch boxes from the residence of workers in the morning (9am) and delivers the same to the persons or officers who do jobs in towns or cities area (12.30pm) and at the late afternoon (4pm), they deposit the empty lunch boxes to the respective customer's residence. It is a case study of unemployed, illiterate and poor people, who by virtue of their hard labour and symbolic communication pattern are able to get Rs. 5000-6000 per month at the upfront investment of roughly 5000 rupees. A cosmopolitan culture of Mumbai customers, satisfying hunger with home cooked food, a hectic journey schedule in the local train creates and maintains the demand of the dabbabala in that particular region. The dabbabala only charge Rs. 150/- to Rs. 300/- per dabba per month depending on the location and collection time. Their association is known as **Mumbai Tiffin Box Suppliers Association** (MTBSA) where almost 4500 dabbabalas are doing their activities. No doubt the market is more tensed in Mumbai particularly in food and beverages segment, still the simplistic model of dabbabala has an indelible mark in business spheres. According to an interview with the daily The Hindu, veteran dabbawala Jayasingh Ghorpade says: "Let the youngsters eat junk food. Our targets are the office and factory workers. Occasionally, they may stop our services and opt to eat at the canteen/restaurant. But in three months or less, they are back".

## Unemployment Changes to Self-Employment

The balanced growth theory is an economic theory pioneered by the economist Ragnar Nurkse (1907–1959). The theory hypothesizes that the government of any underdeveloped country needs to make large investments in a number of industries simultaneously. Actually what happens, in under developed and developing countries, the dependency on agriculture is quite high in comparison to develop one. According to data of 2009-2010, where the contribution of agriculture to the national income in India is 14.6%, the UK is 2%, USA is 3%, Canada is 4% and Australia is 5%. As a result our maximum population in agriculture either faces unemployment due to seasonal nature of agriculture or faces disguise unemployment due to low productivity. In this instance, promoting entrepreneurship through agro industries can change the scenario of rural and agriculture sector along with shifting excess labour from agriculture to industries. As a result, the overall productivity in both agriculture and industry will be enhanced and social maladies like unemployment can be changed to self-employment.

### Caselet

*Café Coffee Day*

This is an innovative venture of Mr V.G. Siddharth who gives the track to one of the beverages of India i.e. coffee, to a brand name under an ambience of cosmopolitan culture and by the help of information technology. The path was not as simple as it is reflected. Coming from an agriculture based family to an academic domain (P.G. Economic) and then to move in the path of commodity arbitrage and speculation in commodity market, need an enough risk taking ability and enthusiasm. He returned to family estates, after narrowly having lost out on joining the army. He told his father to do business and his father also had given 7,50,000 rupees against a promise that if he fell, then came back to their traditional agriculture without a question. Mr Siddharth invested money in stock market and trading. When he invested money, he always invested in coffee trading due to his background. On the way, he realized that there was a great scope in coffee market when a free market is in reality. He entered coffee marketing being a grower he always confronted with middlemen to market his produce and on the way he always got a very little proportion of profit and maximum amount was taken by the middlemen. After that he established Café Coffee Day (CCD) in two main cities in India i.e. Bangalore and Chennai. Proper location selection, team effort, regular innovation and young target customers took his business to the path of glory. He always thinks, "take the path less travelled, you may get lost in the forest or you may succeed".

## Arrest Migration of Youth and Talent through Social Entrepreneurship

From health to education; and from environment to information, societies always try to march one step ahead as the basic objective of the social system is to enhance overall empowerment among each stakeholder of the society. The basic difference between the entrepreneur and social entrepreneur is that the later always think about the society instead of profit. The society faces poverty and paucity of funds and in the meantime social entrepreneur produces innovative business model to grasp such maladies in a better way.

Now-a-days corporate sector also penetrates the society in a concept of corporate social responsibility. According to new Companies Act (2014), all the companies with a turnover of Rs. 1000 crore or more or networth of Rs. 500 crore or more or profits of Rs. 5 crore or more must invest annually, a minimum 2% of their net profits (calculated on a three year average basis) on CSR activities. The act has defined ten broad areas for CSR activities, mainly covering activities in the rural development space. It may be the already existing innovative concept like that of HLL in the form of Project Shakti where most deprived persons particularly women generate a handsome money through an innovative business ideas or through innovative concept of microfinance by a social entrepreneur Mohammud Yunus where society is integrated in a self-help group.

### Caselet

#### *Grameen Bank*

Before coming to Grameen bank, we should understand the apathy and malady of the society in which this bank concept has been established in Bangladesh. When the country separated from Pakistan in 1971, the citizens of this country were facing abject poverty, were dwelling in overcrowded cities and were facing the devastation of catastrophic weather. No one had enough capital to take loan from the bank. Besides that, banking institutions did not prefer to give loans to the women. At that juncture of weak economy and social sabotage, Prof. Mohammad Yunus, an economist studied the landless underclass of Jobra Village near Chittagong. At that time, Prof. Yunus targeted women as their members as he felt the women were the persons if benefitted then its reflection would be observed throughout the family. In 1976, Prof Yunus started his innovative concept by providing 25 dollar each within 43 villagers and by 1979; he included 500 borrowers in different groups. Hopefully from that initial venture, the concept of Grameen Bank emerged. This is a case study that speaks about welfare of the women in the society, social upliftment, and poverty eradication through micro finance. At the end, Grameen Bank came to sight in 1983.

### It Preserves Culture and Value through Agro-Tourism Enterprise

Agro-tourism is a lucrative sector for the entrepreneurs in these days as it speaks about social harmony and generates aesthetic environment in the country. Kerala is spoken as God's own country. In Maharashtra, rural areas have formed an organization named Maharashtra State Agri and Rural Tourism (MART). There are about 150 Agri-tourism centres in the state working without financial assistance of the government schemes. Rajasthan is one of the preferred ventures to attract the foreign visitors. In Himachal Pradesh, the state government is motivating the rural people to create required facilities in rural areas for tourism purpose.

Agri-tourism can be defined as - a range of activities, services and amenities provided by farmers and rural people to attract tourist to their area in order to generate extra income for their businesses. Question arises, why agro-tourism is the preferred venture of the entrepreneur? The numbers of respondents spoke that it was family based avenues in the rural sector that attract health conscious and curious urban people to come and to spend few days in a completely recreational environment.

**Caselet**

***Farm of happiness: home stay at organic farm***

This is the story of two copreneur who were born and brought up in Mumbai city and have settled in a small village in Konkan named Phungus. The searching of traditional agriculture and the ancient culture is the prime aim of this copreneur, named Rahul and Sampada. Actually they set up an organic farm in aforesaid village and the visitors are attracted for plucking and buying of fruits and vegetables. They represent themselves as urban as you can find the environment of urban modernity combined with rural earthiness in their farm. They try to motivate visitors to stay at their organic farm and make them feel a pleasure of agriculture through experience of transplantation of seedlings, fishing, trekking, bird watching, star grazing and many more. "Grow your own food and feel the pleasure of the place" is the basic objective of this venture.

### It Maintains both Forward and Backward linkages through Agro Industries

The primary sector in any economy is agriculture and the secondary is industries. Both are interlinked in agro industries as the produce of one sector is utilized in another. But thing is that it was not at the same pace before the independence. The swadeshi call of Mahatma Gandhi replaced machine with man and the major emphasis was given on khadi and village industries as it spoke about our culture and austerity against the communalism. The picture somewhat changed after 1980 and the agro industries found some momentum in the form of food industries. After liberalization, globalization, privatization (LPG) era, foreign direct investment found some lucrative ventures in the agro industrial sector. The taste of burgeoning young and nuclear family found convenience to get everything under a single roof and that latent demand motivated number of entrepreneurs to invest in different ventures. As we are more concerned about wastages, productivity, information and organized institutions, so to get entrepreneur in agro industrial sector is aboon for the society. Agro industries are different types like agro manufacturing, agro service etc. An example is Agri Clinics and Agri Business Centres (AC&ABCS) – a scheme supported by MANAGE and NABARD. Such schemes not only benefit the entrepreneur but also provide in time information to the primary producer. It is very difficult to consume some of the products like paddy, barley in its whole form and the perishability of some of the products like milk, meat, fruits, vegetables, fish is for few hours only. These agro industries compel entrepreneur to search some profit avenue in the form of processing and that processed product can be preserved for a long time and consumed throughout length and breadth of the country.

**Caselet**

***Manipur entrepreneurs tap potential for agro industries***

Manipur is a state that has immense potentiality for agro industries and business environment is such that it attracts entrepreneurs to Manipur. Thangjam Joykumar Singh, a young entrepreneur of the state of Manipur invested in food and beverage production unit and created job opportunities for a large chunk of people. Mr. Singh advertised different food items in the brand name of 'Likla', which literally means 'dew drops'. This agro processing unit is the source of employment for local youth as it employed about 200 local youth. The company also prepared and packaged drinking water in the same brand name. The unit prepares beverage items in the category of orange, mango, pineapple and passion fruit squash. In the era of unemployment and illiteracy and social maladies like extortion, it is very difficult to find a job. Such types of agro industrial units create social harmony and ultimately add positivity to the direction of growth.

### It Increases Social and National Integration through Women Enterprise

Prime Minister Pandit Jawaharlal Nehru stated- You can tell the condition of a nation by looking at the status of the women. The women are so much important from entrepreneurship point of view, is because of their submissive nature, internal knowledge and tacit experience. Entrepreneurship is the instrument that gives a platform in the direction of innovation, organizing and profit. When women are in the bracket of entrepreneurship then a wholesome approach of empowerment will be possible. These days government promotes backyard poultry farming, mushroom cultivation, bee keeping, and floriculture like activities through women as it is quite approachable for them inspite of their limitations. There a very good success story of Mrs Thezanofrom Sethikema, a woman entrepreneur in Nagaland village who started an enterprise with 400 backyard poultry birds and became successful with a net benefit of Rs 39,470/- in the first year. The scientific skill has been provided to her through Indian Council of Agricultural Research (ICAR). The story gives a lesson that with a sheer determination and proper utilization of local resources one can become a very good entrepreneur with proper integration and focused approach.

**Caselet**

***Mrs Bector's Cremica- A journey of an entrepreneur who had never planned***

The story of an entrepreneur named as Rajni Bector who left her birth place Lahore after partition, did her schooling at Delhi, married into a business family in Ludhiana and continued as a typical housewife life i.e. nurturing her three sons for a better future. She found her life alone when her sons left her and went for higher studies. In an eagerness to learn how to cook a delicious food, she took admission in a course in baking in Punjab Agricultural University (PAU). She always searched occasions to show her skill in baking to her friends and relatives. Day by day she became famous among her near and dear ones, and some of her friends gave her advice to start a business in creams, cakes and cookies. In initial days she always put her effort to satisfy the customers, even if she sold her products below the cost of production. Then upon advice of her husband, she changed her home based catering service to a business model which later came to be known by the name of Cremica in 1978. From catering to business, the journey was tough and it was even tougher when their family business collapsed in 1990 due to terrorism. Then all of her family resources from man to material came to a single platform of Cremica and then their business found substances to spread everywhere with name and fame. This is a story of a woman entrepreneur who stepped in to her kitchen as well as stepped out of the boundary walls and shows that life can change its course at any point of time.

### It Provides both Salary Substitution Form (laundry) and Lifestyle Substitution Form (USA)

The growth of a country is reflected in her gross domestic product (GDP) or in her national income. In earlier times the service of women were never remunerated and could not be accounted in our national income. However, the situation is completely different as today we have a large mass of young generation, our family is quite nuclear and both the partners of our family are job seekers. As a result, the traditional home based activities have changed their texture and new business avenues are being created for start-up entrepreneurs. Fast food and fast life creates a psychographic segment for the entrepreneurs in the market that targets the customers by looking at their personality. These days, entrepreneur produces and markets his products with the help of technology as it saves time, money and labour. Dress and address speak a lot that is why Nestle found its market in cosmopolitan market like Mumbai and Bru found its market in Chennai due to her traditional and culture based campaign. It means as India is bestowed with various cultures, sub cultures, religions etc so here entrepreneurs find their business ideas either looking to west (US like advanced countries) or looking to east (Singapore like tiger economies). Entrepreneur acts as a linkage among the markets, creates liaison between the producers and consumers and provides opportunities through continuous innovation and invention.

**Caselet**

***Young entrepreneur and agri-business scenario in India***

- Technology changes the life and also establishes business opportunity even in the arid region of Kutch of Gujarat. Here the name of the entrepreneur is Mr Rahul Gala who starts his day by logging in his computer and feeding the data in it. Then everything is managed by computer from irrigation to fertigation. In his farm he grows export quality dates (particularly barhi variety) along with mangoes and exports to them to Australia and Dubai under the brand name of Golden Barhi. This business suits to him as his degree is in horticulture (Australia). He also tied-up with seven super varieties in Mumbai and invested almost Rs. six crore in cold storage facilities to keep his produce fresh for further export. These days agriculture offers a number of opportunities if approached with dynamism and professionalism. This is again proved by this case study.
- Similarly another case study is also based on an agri-entrepreneur. Mr Kumar Patel, resident of Kutched village in Valsad, Gujarat. As this region is dominated by NRIs (Non Resident Indians), to encash the demand of this segment, Mr Kumar Patel visualized business opportunities in rose cultivation under the brand name Best Roses. He exports his produce to USA, Japan, Holland, Europe and Dubai and he also produces roses in hydroponic environment which was first in India particularly in this segment. The earning capacity of the laborers increased upto 40% though they were illiterate and could not understand the conventional method of cultivation of sugarcane and paddy in that region. This case study states that entrepreneur and his visionary outlook agriculture to agri-business through value addition.

To summarize the role of entrepreneur in economic significance in one line is that it creates stability, flexibility and a sense of identity in both local and global scenarios through innovation, integration and inclusiveness.

## Challenges in Agro-Entrepreneurship

When we discuss about any challenges, its major focus is on time, cost (investment cost) and resources. If we have very less time, then we capitalize our goal by adding some extra cost and when we have time flexibility then we will plan our investment processes accordingly. An entrepreneur always dwells in between time and cost to realize benefits. Otherwise his project may fail due to time overrun or cost overrun.

The problem to tackle investment cost is a daunting task for today's agropreneur. In today's scenario our major credit flows in agriculture depend on informal sector due to its flexibility in terms of collateral and documentary simplicity. Besides that, though major agriculture produce is perishable, a major amount of cost is required for working capital management unlike other group of investments like housing or car loan. The lending institutions should understand this peculiar nature of investment to generate some substantial level of investment in agro- entrepreneurship.

People always say that access to finance/investment is the key issue for entrepreneurs. But through various perceptions it is proved that it is very difficult to approach financer than finance. That's why these days we emphasize more on financial inclusion. Financial inclusion or inclusive financing is the delivery of financial services at affordable costs to sections of disadvantaged and low-income segments of society, in contrast to financial exclusion where those services are not available or affordable. Customer relationship management and partner relationship management are two focus points to attract and motivate the agro-entrepreneurs in to agri-business.

In today's business, you sell your produce today otherwise it could be too old tomorrow. It is easier said than done, as lack of proper infrastructure like roads, electricity (energy), cold chain etc, creates the bottlenecks for smooth flow of agricultural produce to the demand area. Till date the use of electricity for agriculture in India is maximum up to 18%, and modern agriculture also relies upon machinery that runs on gasoline and diesel fuel (e.g., tractors and combines) and different equipment that use electricity (e.g., lights, pumps, fans, etc.). No doubt, we have been able to motivate the primary producers to use micro irrigation systems in states like Gujarat and Maharashtra but still our agriculture in many states like Odisha, Bihar is at the mercy of monsoons.

The journey of an entrepreneur is similar to path of roses, where you get the thorns in the waft of rose. Sheer determination, continuous dedication and unstoppable discipline are necessary for marching ahead on the entrepreneurial platform. Someone also explains the journey of entrepreneur as the ride on the back of the tiger. You enjoy the drive but you cannot retrieve back. That's why, scanning of the environment by the entrepreneur is very necessary. If you have more information you will be on the driver seat and to get the information you should know different tools and techniques and you have to invest in them on a continuous basis.

In India, gender inequality is a major problem in the growth of entrepreneurship. Even in financial matters male interference is seen in documentation and investment. This is the reason; it is very difficult for the financial institutions to credit loan in the name of a female beneficiary. Besides that male dominating culture restricts the interest of women in entrepreneurship. It is also observed that youth is facing discrimination in the ground of experience that comes in later part of the life; as a result they separate themselves from positivity and optimism.

## Charms of Being an Agro-Entrepreneur

Entrepreneurship is a complete package of passion, perseverance, success and satisfaction. Living a life with thrill is the hallmark of an entrepreneur. Entrepreneur creates an idea, works on it till he finds his goal and objectives. In today's stereotypic environment, it is very difficult to balance family, job, hobby, health and spirituality at a single point. One parameter always dominates over the other and everyone's life goes on without any complete solution. At this juncture, entrepreneurship comes as a panacea to stabilize the life in an order and shows a new route to all the individuals to prepare themselves as an entrepreneur. Entrepreneurship makes them to feel proud to be a leader and controller of their destiny.

Business comes from an idea and idea comes from a mind. No doubt government works on policy formulation, implementation and its control; private players work on profit maximization or cost reduction or both; but at the end, no organization can grow without new ideas as business in today's world faces cut throat competition. To search a business opportunity from the market is both an art and endeavor for the businessman and these days to feed more than 121 crore population is a daunting task for the government, private players and an entrepreneur. This paradigm builds the concept of agro-entrepreneurship which speaks about toil of soil and its paramount importance in Indian culture, as farming offers an eternal bondage with our long lasting primitive civilization.

When agriculture turns to agri-business then traditional farmer transforms into commercial entrepreneur. Agro-entrepreneurship sets up avenues for profit making for builders (in the form of supermarket), IT personnel (in the form of design of package, software build up and database maintenance for agriculture) and retired military personnel (in the form of family farming and farm house establishment) in their recession times and gives them stability and soundness in their life as ever before. Now both quality and quantity of natural resources like land, water and air recede every day and their effective and efficient utilization are necessary to generate maximum productivity. As supply of resources is constant and demand for food is increasing day by day that's why agricultural production and its proper distribution to the ultimate consumer may be a direction for an agro-entrepreneur to invest on it for profit escalation. A blooming segment of entrepreneurs are evolving into ecopreneurs, passionately creating businesses that go beyond building just profit, but reflect a desire to leave the world a better place and redefines success not by bank account size but by life's tangibles, health, wellness, meaningful work, vibrant community life and health. An agro-entrepreneur has many options in agriculture as he redefines his business in his leisure time and communicates many indelible feelings with his family and can be capable to inculcate skills of farming to his kith and kin which descends from generation to generation. It means to start a life as an agro-entrepreneur is a sustainable way of creating one's future.

Last but not the least; agriculture is a field in which we never find recession. It is the venture in which many areas starting from production to processing and marketing to consultancy are in its nascent stage and their complete exploitation is yet to happen. In the era of sustainability, food security, food safety and hygiene, family farming, climate resilience and agri-business, the role of agro-entrepreneur is quite challenging and demanding as well. Hopefully when we transform the mindset of our farmer to that of an entrepreneur and when we find maximum dependency on agri-business in the form of agro-entrepreneurs, a day will come when agriculture turns to be the largest industry in this world. Then agro entrepreneurs can proudly justify themselves as industrialists who build the nation and generate everlasting success and satisfaction for themselves.

■■■

*Chapter 2*

# ENTREPRENEURSHIP OPPORTUNITIES IN AGRI BIOTECH SECTOR

**Mounil Mankad**

*Research Associate, Plant Tissue Culture Lab., Anand Agricultural University, Anand*

## INTRODUCTION

In the coming years, technology will have a significant impact on society, industry and the consumer, both at national and international levels. Among various technologies, information and biotechnology will certainly have a bigger impact in addressing the global issues as well as have great possibilities for new generation to start- up their own small business and become a leader in entrepreneurship in these fields. As far as biotechnology is concerned, it has already started to revolutionize science, health, energy, resources, environment, consumer products and manufacturing processes. To address global issues using biotechnology, it is very much essential for the successful translation of laboratory research and development into commercialized new products, services and systems.

Biotechnology is the application of science and technology to living organisms and products of living organisms, to produce knowledge, goods and services. Biotechnological interventions have played a significant role in understanding the genomic information and identified novel genes for targeted genetic engineering; development of new therapeutics and personalized medicines; development of a number of biological based products such as bioplastics, biocatalysts and advanced biofuels. The major impact of biotechnological interventions is in the field of agriculture which includes addressing issues pertaining to crops especially enhancement in yield providing abiotic and biotic resistance, identification of new quantitative trait loci for their utilization in breeding programmes etc.

Looking at the immense possibilities in the biotechnology field at commercial level and its possible exploitation in developing new products, the biotechnology industry faces many barriers to success, most important of which are those that affect the development of appropriate research and technology transfer capability, including access to funding, shortage of skills and regulatory issues. In addition to several regulatory and ethical issues, the major barrier is the expenditure in development of product and its commercialization through various clinical and on field trials in medical and agriculture biotechnologies, respectively.

## Influences Affecting Adoption

In the next ten or twenty years, cutting edge technology like biotechnology would be able to meet the consumer expectations in providing new drugs with better efficiency, upgrading already existing manufacturing practices like overcoming downstream problems of recovery etc. In achieving these goals, various factors have been identified that will influence the successful adoption, execution and utilization of biotechnological interventions. Some of them are listed below.

1. **Market-pull commercialization:** The need for demand driven commercialization strategies that focus on developing new products and services to address existing problems and challenges.
2. **Absorptive capacity:** The ability to recognize valuable new enabling technologies and applying them to commercial ends.
3. **Government support for research and enabling technologies:** The need for continued government support for research, infrastructure, training, and collaboration.
4. **Convergence of technologies:** Increased interdisciplinary — particularly convergence with the cognitive sciences — leads to challenges around regulation, funding requirements and skills development.
5. **Collaboration between research and industry sectors:** The need to develop strong networks between research and industry sectors to ensure public acceptance of new technologies.
6. **Incremental, radical and transformational innovation:** Speedy innovation is a significant challenge for enabling technology based companies who find issues such as commercialization and scalability leading to long research and development (R&D) cycles.
7. **Product innovation versus market innovation:** Business model innovation can allow firms to enter into new market spaces.
8. **Knowledge of enabling technologies:** Promoting public awareness of enabling technologies and ensuring democratic engagement with the wider community in decision making about technology development.
9. **Regulatory environment:** The regulatory environment can influence the direction of research into enabling technologies, the type of research that is commercially viable, and the costs of research and development.
10. **Intellectual property rights:** Protection of intellectual property is particularly important for enabling high end technology based investments because the research is capital intensive with long and costly lead times.
11. **Ethical considerations:** Technological advancements are bounded by ethical issues, such as how we can or should change ourselves and our environments leaving behind many traditional unsustainable practices. Recent advances in molecular biology and genomics research are driving the revolutionary change and growth in the biotechnology industry. The major activities, products and services in the biotechnology industry are currently:

a. Agricultural biotechnology research and products;
b. Biofuel research and products;
c. Bioinformatics;
d. Chemical, environmental and other research and products;
e. Diagnostics;
f. Human therapeutics; and
g. Reagents and other active molecules.

This section surveys the landscape of biotechnologies, segmented by the following areas of application- conventional and non-conventional biotechnologies. Given that incremental change in current biotechnologies plays an important role in future innovation, this section describes some technologies that have already been commercialized. The source for basic and infrastructural requirements for organic farming has been referred and modified as per the current market scenario wherever applicable from the Entrepreneurship Development Programme in Biotechnology, 2007 conducted by Biotech Consortium India Limited (BCIL), New Delhi and Small Industries Service Institute, Mumbai.

## INTRODUCTION TO ORGANIC FARMING

India is mainly an agricultural country, where agriculture contributes to about 13.7% in gross domestic product (GDP) and supports over 58% of nation's population for livelihood (GoI, 2010).The concern about the health issues due to increased and non-judicious usage of chemical fertilizers has opened upon a new alternative method of producing food grains known as **organic farming**. The food produced using organic farming practices has already gained a niche position in Indian and global markets. Organic food products have not only reduced the usage of fertilizers and pesticides but also fetch a good return to the producer too.

### The Key Advantages of Organic Farming

1. It is self-sustaining, socially and ecologically superior over the conventional farming.
2. It makes use of cost-effective management practices involving the use of farm inputs produced within the farm.
3. Organic certified products with certification marks such as, **India Organic** fetch a premium of around 30% over the conventional products in the domestic as well as global markets.
4. It is less expensive in the long run.
5. It is environment friendly as there is no pollution of soil and water.
6. It enriches the soil and the local ecology.
7. The produce is free of contamination from chemical residues possess better taste, flavor and nutritional value.

### Regulations for Organic Farming

All the organically derived products require authentication in the form of a certificate claiming that the product is organic and has been produced without utilizing synthetic inputs in terms of chemical applications. The certification process involves converting the entire growing area into organic area within a period of three years. Hence, farmers or growers who wish to go for organic management need to wait for upto three years under certification procedures for assuring that the area is completely chemical free. The concept of organic standards extends from farm input use and land preparation to the post-harvest management, storage, processing, packaging, transportation as well as handling. This involves careful monitoring, inspection and certification at all stages right from sowing/planting till the product reaches the consumer market.

In India, the Director General of Foreign Trade (DGFT), New Delhi, permits the export of organic products provided that these are produced, processed and packed under a valid organic certificate issued by a certification agency accredited by an accreditation agency designated by the Government of India. The Government of India has recognized Tamil Nadu Organic Certification Department, Agricultural and Processed Food Products Export Development Authority (APEDA), Spice Board, Ministry of Commerce and Industry, Coffee Board and Tea Board for the certification purpose.

### Market Potential

Currently, India ranks 10th in terms of cultivable area under certified organic cultivation. The certified area includes 15% cultivable area with 0.72 million hectare and rest 85% (3.99 million hectare) is forest and wild area for collection of minor forest produces. The total area under organic certification is 4.72 million hectares (2013-14) *(http://apeda.gov.in/ apeda/website /organic/ Organic_Products.htm)*

### Production

India produced around 1.24 million MT of certified organic products which includes all varieties of almost all food products namely sugarcane, cotton, oil seeds, basmati rice, pulses, spices, tea, fruits, dry fruits, vegetables, coffee and their value added products. The production is not limited to the edible sector, but also includes organic cotton fiber, functional food products etc. Among all the states, Madhya Pradesh has covered largest area under organic certification followed by Himachal Pradesh and Rajasthan.

### Exports

India exported 135 products in 2013-2014 with the total volume of 194088 MT including 16322 MT organic textiles. The organic agri-export realization was around 403 million US$ including 183 million US$ organic textiles registering a 7.73% growth over the previous year. Organic products are exported to US, European Union, Canada, Switzerland, Australia, New Zealand, South East Asian countries, Middle East, South Africa etc.

Oil seeds - Soybean (70%) leader among the products exported followed by cereals & millets other than Basmati (6%), processed food products (5%), Basmati rice (4%), sugar (3%), tea (2%), pulses and lentils (1%), dry fruits (1%), spices (1%) and others.

The demand for organic products over conventional products is also growing steadily in India. This concept is spreading widely in the metros where organic products are being sold in exclusive organic shops eg. Organic Love in Guwahati, Sresta in Hyderabad etc.

**Farm Designing and Selection of Crops**

The most important criteria for farm designing includes selection of the crop which further depends on optimum diversity of suitable species so as to meet the soil conditions to minimize nutrient losses. Border plantation with higher biomass producing crops like leguminous species will also help in generating sufficient fodder for cattle and for preparation of compost etc. For getting year round production, mixed or inter culturing of suitable crops can be done which will help in generating enough revenue from the same soil. The selection of varieties is dependent upon ultimate consumer preferences; however, in general local varieties should be given first preference.

**Crop Protection**

Plants are exposed to various bacterial and fungal diseases and also with a number of different types of insects throughout their life cycle. Hence, it is very much necessary to control the spread of diseases and pest attacks which otherwise leads to severe yield loss leading to reduction in production. However, under organic cultivation, the use of routine chemicals and sprays for control of these microbial and pest attacks are not allowed. Hence, control measures include rotation of crops, green manuring, rearing and maintaining natural enemies of predator organisms, spraying of botanical extracts of plants having antimicrobial property, biofertilizers, biopesticides etc. A specialized mixture of cow dung, urine, milk curd and ghee (*panchgaviya*) can be used in pest management.

**Project Details**

The project details with all necessary requirements to execute organic farming have been discussed under this section. The proposed project covers 10 hectare area of organic cultivation with following cropping pattern.

**Cropping Period**

1. Rice: May-November
2. Wheat: November-April
3. Pulses: April-May
4. Vegetables: April-May
5. Seasonal vegetables can also be grown throughout the year.

## Project Cost

### A. Fixed Assets (Table 2.1)

**Table 2.1:** Fixed assets required for organic farming in one hectare

| S.N. | Particulars | Unit Cost (Rs.) | Units | Cost (Rs. in lakhs) |
|---|---|---|---|---|
| 1 | Land rent | 20,000/ha | 10 | 2.00 |
| 2 | Shed/store house for produce under conversion | 250/sqm | 220 | 0.55 |
| 3 | Cattle (Local breed) | 8,000/cow | 10 | 0.80 |
| 4 | Power tiller | 1,25,000 | 1 | 1.25 |
| 5 | Training expenses (10 farmers for one week) | 35,000/annum | 1 year | 0.35 |
| | | | Total | **4.95** |

### B. Working capital requirement

I. **Raw material (Table 2.2)**

**Table 2.2:** Raw materials required for organic farming

| Seeds | Kg/ha | Rs. / kg | Cost for 10 ha. (Rs.) |
|---|---|---|---|
| Rice | 100 | 20 | 20000 |
| Wheat | 100 | 20 | 20000 |
| Vegetables | -- | -- | 7500 |
| Pulses | 50 | 20 | 10000 |
| | | Total | 57,500.00 |

**Note:** If seeds are produced on – farm, the necessity to procure them does not rise.

II. **Manpower**

Labour for land preparation, composting, harvesting, driving and storage etc. – Rs. 150/person/day. Therefore, approximately Rs. 1,00,000.00 will be required towards hiring manpower.

III. **Cost of milling** Rs. 45,000

IV. **Utilities (Power 2000 unit @ Rs. 7/unit)** Rs. 14,000

V. **Certification (per annum)** Rs.30,000
(this includes certification of the farm upto 50 ha area)

VI. **Contingencies expenses (marketing, maintenance etc.)** Rs. 75,000

VII. **Recurring expenses (per annum)**

Raw material Rs. 57,500

Manpower Rs. 1,00,000

Cost of milling Rs. 45,000

| | |
|---|---|
| Utilities | Rs. 14,000 |
| Certification | Rs. 30,000 |
| Contingencies | Rs 75,000 |
| Total | Rs. 3,21,500.00 |

**C. Capital investment**

| | |
|---|---|
| I. **Fixed assets** | Rs. 4.95 (in lakhs) |
| II. **Recurring cost (per annum)** | Rs. 3.21 (in lakhs) |
| Total | Rs. 8.16 (in lakhs) |

**D. Means of finance**

| | |
|---|---|
| I. Equity | Rs. 3.26 (in lakhs) |
| II. Debt | Rs. 4.89 (in lakhs) |
| Total | Rs. 8.16 (in lakhs) |
| Debt : Equity | 3:2 |
| Interest | 16% p.a. |

**E. Finance analysis**

I. Cost of production

| | |
|---|---|
| Recurring cost (per annum) | Rs. 3.21 (in lakhs) |
| Depreciation (@10%) | Rs. 0.19 (in lakhs) |
| Interest (@16% per annum) | Rs. 0.79 (in lakhs) |
| Total | Rs. 4.19 (in lakhs) |

II. Turn over **(Table 2.3)**

**Table 2.3:** Turnover in organic farming

| Seeds | Yield (T/ha) | Selling price (Rs./kg) | Total (10 ha.) (Rs. in lakhs) |
|---|---|---|---|
| Rice | 2.0 | 80 | 2.80 |
| Wheat | 2.0 | 11 | 2.20 |
| Vegetable | 7.0 | 7 | 4.90 |
| Pulses | 1.5 | 10 | 1.50 |
| | | Total | **11.40** |

**III.** Profitability (turn over - cost of production) (Table 2.4)

| | |
|---|---|
| Net profit | Rs. 8.28 lakhs |

**Table 2.4:** SWOT analysis for organic farming

| **Strength** | **Weakness** |
|---|---|
| • Comparative advantage in organic food production<br>• Safe food<br>• Low cost of production<br>• High quality and improved nutrition<br>• Improved soil health<br>• Premium prices<br>• Environmental sustainability<br>• High water-use efficiency<br>• Enabling Government policies<br>• Preserves traditional varieties/ species and high self-life | • Productivity gaps<br>• Lack of established markets<br>• Poor quality management in production and processing<br>• Less incentives from Government<br>• Low R&D investments on Organic farming research<br>• Organic market buyers/consumers driven market<br>• Lack of strategy for development of organic market<br>• Disjointed producers, processors and traders<br>• Adulteration and poor quality of organic inputs<br>• Large number of small farms with weak organizational building<br>• Intensive in nature and high labor costs |
| • Opportunities<br>• Big and growing market potential<br>• Growing purchasing power of consumers<br>• Growing health awareness<br>• Reduce heavy subsidies on food and fertilizers<br>• Control the nitrate losses and $CO_2$ emissions<br>• Earn high export earnings | • Threats<br>• Costly and complex organic certification process<br>• Lack of infrastructure facilities (like labs) and certification bodies<br>• Only export regulated organic market<br>• Low awareness about organic inputs<br>• Most of the fields are contiguous and problem of contamination |

## INTRODUCTION TO LIQUID BIOFERTILIZERS

Indian green revolution has played an important role in making India a self-dependent nation in terms of feeding its people. The green revolution could become possible only due to introduction of high yielding varieties (HYVs) of different food crops. However, with the introduction of these HYVs, usage of chemical fertilizers has also increased exponentially leading to ever growing demand of fertilizers. Due to lack of proper guidelines and regulatory framework until the recent decade, the indiscriminate use of chemical fertilizers by the farmers led to environmental pollution, erosion of soil quality and contamination of farm produce and groundwater by chemical fertilizer residues and consequent health hazards.

Nitrate poisoning and extermination of beneficial microflora and microfauna, adversely affecting the physical and chemical structure of the soil were also among the major problems caused by chemical fertilizers. Several reports have been published by various authors stating the possible harmful effects of these chemical fertilizers are seen not only in soil micro-flora but also on humans, cattle and other livestock. Integrated plant nutrient supply (IPNS) systems consisting of a judicious combination of chemical fertilizers, organic matter and biofertilizers ensures optimal nutrient supply to crops, at the same time conserving soil productivity and ecology. In this context, biofertilizers have assumed importance as an essential alternate input to chemical fertilizers. Earlier these were carrier based (powder). Their new variant is liquid formulations popularly known as **liquid biofertilizers** (LBFs). These days a number of state governments are providing subsidies to start liquid biofertilizer plants.

## Types of Liquid Biofertilizers

Liquid biofertilizers are special liquid formulations containing not only the desired microorganisms and their nutrients but also special cell protectants or substances that encourage formation of resting spores and cyst for longer shelf life and tolerance to adverse condition.

Liquid biofertilizers offer a distinct advantage over carrier based formulations of biofertilizers. Liquid biofertilizers have a higher shelf life of minimum one year compared to only six months in carrier based formulations. There is no contamination. It is easy to maintain cell count of $10^8$/ ml. They are easy to store and transport even at higher temperatures. They can be effectively used in drip irrigation.

Microbes localized to soil have several abilities like biological nitrogen fixation, solubilisation of insoluble phosphates in the soil, stimulation of plant growth and control of diseases. The most important and well-known liquid biofertilizers are the following along with their target crop are given in Table 2.5.

**Table 2.5:** Types of biofertilizers, organisms present and target crops

| S. No. | Type of Biofertilizers | Organism present | Target crops |
|---|---|---|---|
| 1 | *Rhizobium* | Bacteria | Leguminous crops (Pulses, oilseeds, fodder) |
| 2 | *Azospirillum* | Bacteria | Rice, sugarcane |
| 3 | *Azotobacter* | Bacteria | Wheat, maize cotton, sugarcane |
| 4 | Phosphate Solubilizing Bacteria (PSB) | Bacteria | All crops |

*Rhizobium* and the phosphate solubilizing bacteria (PSBs) are the two major bacterial groups which play an important role in fixing atmospheric nitrogen and solubilizing the insoluble phosphate, respectively. These two elements, nitrogen and phosphate are the two major nutrients which are required by the plants throughout their life cycle. Hence, chemical fertilizers providing nitrogen and phosphate are extremely utilized leading to increase in alkalinity of soil. However, these bacterial groups can fix and solubilize the nitrogen and phosphate without creating environmental problems.

## Market and Major Players

The growth rate of liquid biofertilizers is around 20% per year which is expected to continue due to increased awareness in people, farmers and growers regarding health benefits, and issues concerned with the chemical fertilizers. There is a great scope of converting the area under carrier based biofertilizers to liquid biofertilizers and area under chemical fertilizers to liquid biofertilizers. The present domestic consumption is less than 3% of the potential market. As liquid biofertilizers are organic in nature, they can be used as an effective input in organic farming (Table 2.6). Area under organic farming in India is increasing for reasons similar to those cited above.

**Table 2.6:** Profile of indicative players in liquid biofertilizers in India

| S. No. | Name of the player | Formulation (Trade name) | Strain (General) | Category |
|---|---|---|---|---|
| 1. | Kumar Krishimitra Bio-products (I) Pvt. Ltd., Pune | Bioplin | *Azotobacter* | N& P Supplement |
| | | Vitormone | *Azotobacter* | N Supplement |
| | | Phosfert | *Azotobacter* & *Bacillus* | N& P Supplement |
| | | Milastin | *Azotobacter* | N Supplement (antifungal strains of *Azotobacter*) |
| 2. | Hindustan Antibiotics Limited, Pimpri (Pune) | Azotomeal | *Azotobacter* | N Supplement |
| | | Phosphomeal | PSB | P Supplement |
| 3. | Agriland Biotech, Vadodara | Agriland *Azotobacter* Liquid | *Azotobacter* | N Supplement |
| | | Agriland PSM Liquid | *Bacillus coagulans* + *Torulaspora globosa* (PSY) | P Supplement |
| | | Biofield Liquid | *Azotobacter chroococcum, Bacillus coagulans, Torulaspora globosa* (PSY) | N & P Supplement |
| 4. | Neesa Agritech Private Limited, Ahmedabad | Vardaan (Cambay's) | *Rhizobium, Azotobacter Azospirillum* | N & P Supplement |
| 5. | Anand Agricultural University, Anand | Anubhav *Azotobacter* | *Azotobacter chroococcum* | N Supplement |
| | | Anubhav *Azospirillum* | *Azospirillum lipoferum* | |
| | | Anubhav PSB | *Bacillus coagulans* | P Supplement |
| 6. | Pramukh Agri-clinic, Surat | Fungbact Nitro | *Azotobacter Sp., Azospirilum Sp.* and *Acetobacter Sp.* | N Supplement |

| | | | | |
|---|---|---|---|---|
| | | Fungbact Phospho | *Bacillus megatharium and Pseudomonas putida* | P Supplement |
| | | Fungbact Potash | Potash mobilizing bacteria, *Fraturia aurentia* | K Supplement |
| | | Fungbact Patron | VAM | P supplement |
| 7. | Tari Biotech, Thanzavur | Azosfer | *Azospirillum* | N Supplement |
| | | Azotofer | *Azotobacter* | N Supplement |
| | | Kmofer | *Frateuria aurantia* | K Mobilizer |
| | | Phosphofer | PSB | P Supplement |
| | | Rhizofer | *Rhizobium* | N Supplement |
| 8. | Aumgene Biotech, Surat | Sugarzyme | *Gluconacetobacter diazotrophicus* | N Supplement |
| | | Biobact | *Azotobacter croococcum* | N Supplement |
| | | Biosol P | *Bacillus megaterium* | P Supplement |
| | | Biofix | *Azospirillum* | N Supplement |
| 9. | Prabhat Fertilizer and Chemical Works, Karnal | Prabhaphos | *Bacillus megaterium* var *phosphaticum* | P Supplement |
| | | Prabhaazos | *Azospirillum* | N Supplement |
| | | Prabhapotash | *Frateuria aurentia* | K Supplement |
| | | Prabharhizo | *Rhizobium* | N Supplement |
| | | Prabhaazoto | *Azotobacter* | N Supplement |
| 10. | CORDET, IFFCO, Kasturinagar, Kalol, Gandhinagar | Iffco Biogreen Azotobacter | *Azotobacter chroococcum* | N Supplement |
| | | IFFCO Bio-green PSB culture | *Bacillus coagulans* | P supplement |

## Application Method

Liquid biofertilizers are useful for all major cereal crops, millets, foliage crops, sugarcane, vegetables, tuber crops, horticultural and plantation crops as well as in floriculture. The major biofertilizers being used are *Azotobacter, Azospirillum, Rhizobium* and phosphobacteria.

## Seed Treatment

Seed treatment is the most common method adopted for all types of inoculants particularly in vegetables, cereals and millets etc. The seed treatment is effective and economical. For small quantity of seeds (up to 5 kg quantity) the coating can be done in a plastic bag. For this purpose, a plastic bag having size (21" x 10") or big size can be used. The bag should be squeezed for 2 minutes or more until all the seeds are uniformly wetted. Then bag is opened, inflated again and shaken gently. Stop shaking after each seed gets a uniform layer of culture coating. The bag is opened and the seed is dried under the shade for 20-30 minutes. For large amount of seeds coating can be done in a bucket and inoculants can be mixed directly with hand. Seed treatment with *Azotobacter* and *Azospirillum*, along with PSM can be done.

Seed treatment can be done with any of two or more bacteria. There is no side (antagonistic) effect. The important thing that has to be kept in mind is seeds must be coated first with *Azotobacter* or *Azospirillum*. When each seed gets a layer of above bacteria then PSM inoculants has to be coated as outer layer. This method will provide maximum number of each bacterium required for better results. Treatments of seed with any two bacteria will not provide maximum number of bacteria on individual seed.

1. **Seedling root dipping**: For application of *Azotobacter/ Azospirillum/* PSM on paddy transplanting/vegetable crops this method is used. The required quantity of *Azotobacter/Azospirillum/*PSM has to be mixed with 5-10 liters of water at one corner of the field and the roots of seedlings have to be dipped for a minimum of half-an-hour before transplantation.
2. **Drip irrigation** or **furrow application with compost** or **organic manure**: One liter culture per hectare to be mixed with 50-60 kg soil or compost and applied in furrow before sowing. For drip irrigation, 1 lit per hectare to be in corporate in 200 liter water tank.
3. **Foliar application:** Take 3-5 ml culture; mix with 1 liter of water and spray on leaves during morning or evening.

**Production Process Description**

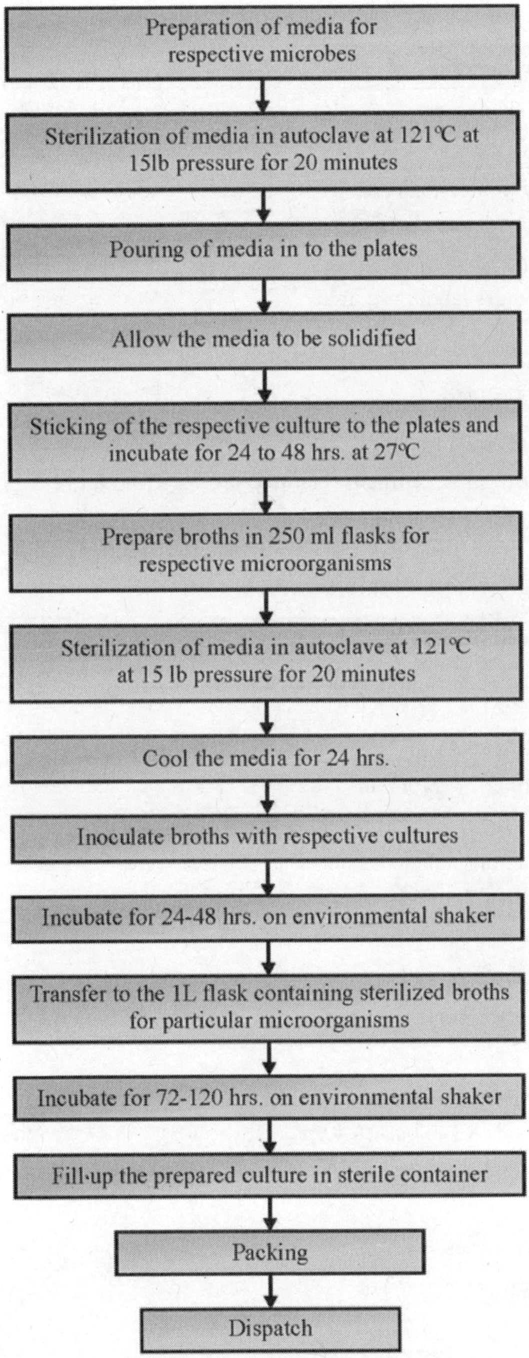


**Figure 2.1:** Process flow for manufacture of liquid biofertilizers

## Project Details

**Table 2.7:** Production capacity for liquid biofertilizers

| S.N. | Details of production | |
|---|---|---|
| 1 | Total production capacity | 5 to 7 Lakh liter |
| 2 | No. of working days | 300 days |
| 3 | Average daily production capacity | 1500 to 2500 bottles /liters (approx.) |
| 4 | No. of shift | 1 (8 hours) + overtime till late evening as and when required. |
| 5 | No. of daily wages labour required | 4 to 6 @ 500 bottle/labor<br>From labor contract and trained in pharmacy sector to work hygienically |
| 6 | Requirement of mother culture | 10 % |
| 7 | Requirement of broth per day | 240 to 250 lit |
| 8 | Requirement of RO or distilled water per day | 2500 liter |
| 9 | Time required for sterilization broth | 3 to 6 hr. |

**Table 2.8:** Costing for liquid biofertilizers production

| S.N. | Item Required (For 7 lakh lit / annum) | Rate (₹) | Annual cost (Lakh) |
|---|---|---|---|
| A | **Fixed cost** | | |
| | Land | | **Variable** |
| | Building = $400m^2$ @ $7000/m^2$ | - | 28.00 |
| | Equipments | - | 26.45 |
| | Glass wares & miscellaneous items | | 3.00 |
| | Technology cost | | 15.0 |
| | Total | | **72.45** |
| B | **Variable cost** | | |
| | (a) Raw materials cost | | |
| | Chemicals | 1.00/lit | 7.00 |
| | Bottle (7 lakh for 1-lit capacity bottle) | 20 ₹/bottle | 140.00 |
| | Carton box (40 bottles,40 liter) | 50 ₹/box | 8.75 |
| | (b) Publicity and marketing cost | 5 ₹/bottle | 35 |
| | Total | | **190.75** |
| C. | **Indirect cost** | | |
| | Salary and wages (manpower) | | 17.28 |
| | Diesel, Maintenance of vehicle | | 3.0 |
| | Office expenses including transport, stationary etc. | | 3.0 |
| | Total | - | **23.28** |

| **Thus, for 7 lakh liter, the production cost will be** | |
|---|---|
| Variable cost | 190.75 |
| Indirect cost | 23.28 |
| Depreciation on fixed asset @ 10 % fixed cost | 7.25 |
| Power consumption ₹. 5.00/unit | 7.00 |
| **GRAND TOTAL** | 228.28 |
| Total cost of production of 7 lakh bottles | 228.28 |
| Therefore, the cost of one bottle | 32.61 |

**Table 2.9:** Different variable cost

| | |
|---|---|
| Risk coverage for unsold Bottles | ₹ 5.00 |
| Coverage for freight & transport cost | ₹ 03.00 |
| Adjustment on variability on input cost | ₹ 01.00 |
| Approx. cost / 1 lit bottle | ₹41.61 (42-approx) |
| Retail sale price /bottle | ₹ 200.00 |
| Commission to wholesaler (5%) | ₹ 10.00 |
| Commission to retailer (15%) | ₹ 30.00 |
| Profit margin /bottle | ₹118.00 |

**Table 2.10:** Equipments and machinery to be bought

| **Equipments** | **Cost/ item** | **Units Reqd.** | **Total ₹** | **Justification** |
|---|---|---|---|---|
| Rotary shaker (36" x 36", 4 flasks having total 30 lit capacity) and out of it one will be having 250 and 500 ml clamp fitted top, heavy duty-1HP motor, max. 250 rpm (with speed controller) | 0.70 | 6+1 | 4.90 | Preparation of starter culture and mass production of biofertilizer. |
| Advance research microscope | 2.00 | 1 | 2.00 | Study of microorganisms |
| Horizontal autoclave 0.9 x 0.9 x 1.8 m | 5.00 | 2 | 10.00 | Sterilization of media, glassware, carrier etc. |
| Autoclave (small) | 0.25 | 2 | 0.50 | Post-sterilization of media |
| Refrigerator 350 lit single door. | 0.30 | 2 | 0.60 | Storage of microorganisms, chemicals |
| Laminar Airflow 180 x 60 x 60 cm | 1.00 | 2 | 2.00 | For providing the microorganisms free atmosphere |
| Air curtain-3ft. | 0.25 | 2 | 0.50 | For providing microorganisms free air. |

| | | | | |
|---|---|---|---|---|
| Electronic balance | 0.25 | 1 | 0.25 | For weighing chemicals etc. |
| RO system with 2000 lt./day storage | 1.50 | 1 | 1.50 | For water requirements |
| pH meter | 0.20 | 1 | 0.20 | For adjustment of pH |

**Table 2.11**: Cost incurred for incubator, colorimeter

| | | | | |
|---|---|---|---|---|
| Digitalized incubator cum oven 18"x18"x18" inner chamber. | 0.60 | 1 | 0.60 | Growth of microorganisms. drying, sterilization of glassware |
| BOD incubator 6 cubic foot | 0.80 | 1 | 0.80 | Growth of microorganisms. |
| Colorimeter | 0.15 | 1 | 0.15 | Quality control |
| Lab centrifuge small | 0.25 | 1 | 0.25 | Quality control |
| Colony counter | 0.15 | 1 | 0.15 | Quality control |
| Air conditioner 1.5 ton with heating facility | 0.60 | 3 | 1.80 | For providing controlled conditions (cooling and heating facility) in laboratories |
| Portable Genset | 3.00 | 1 | 3.00 | For providing electricity during power failure |
| Tullu pump 0.5 HP | 0.25 | 1 | 0.25 | For handling of RO water |
| Other small equipment's costing less than Rs. 10,000/- | 1.0 | - | 1.0 | Trolleys, mixer grinder, hand mixer, water storage tanks food grade and top pan-plastic balance etc. |
| Bottling unit mixing tank with1HP motor. | 1.0 | 2 | 2.0 | Design will be provided. |
| Total | | | 32.45 | |

**Table 2.12:** Manpower requirement

| S.N. | Services | Fixed Pay @ rates ₹ | Post | Year 1 cost (Rs. Lakhs) |
|---|---|---|---|---|
| 1 | Microbiologist (M.Sc. or Ph.D.) | 60,000 | 1 | 7.20 |
| 2 | Assistant Microbiologist (M.Sc. or B.Sc.) | 30000 | 2 | 7.20 |
| 3 | Skilled (2) and office worker (2) | 10,000 | 4 | 4.80 |
| 4 | Labours for cleaning, media, filling, packing, transport | 3500 | 4 | 1.68 |
| Total | | | | **17.28** |

**Table 2.13:** Plant furnishing

| Sr. No. | Name of unit | Size (Sq. m.) | Facilities required |
|---|---|---|---|
| **(A) Inoculums wing** | | | |
| 1. | Storage room | 35 | Racks, cabinets etc. |
| 2. | Media preparation room | 30 | Vertical autoclave, pH meter, distillation plant, deionizer, oven, sink, lab. bench etc. |

| | | | |
|---|---|---|---|
| 3 | Bacteriological room | 30 | Incubator, fridge |
| 4. | Fermentation room | 40 | Fomenters, rotary shakers, lab bench, air conditioner etc. |
| 5. | Quality control room | 30 | Lab bench, sink, microscope, pH meter, colony counter, spectrophotometer, balance |
| 6. | Growth room | 30 | Slotted angle rack with illumination, Leonard jars, air conditioner, timer etc. |
| **Sub Total** | | **195** | |
| **(B) Dilution and Packaging wing** | | | |
| 7. | RO system | 20 | RO machine and cooling system |
| 8. | Curing room | 30 | Air conditioners temp. & humidity controller, corrugated boxes etc. |
| 9. | Packing room and storage of Bottle | 30 | |
| **Sub Total** | | **80** | |
| **(C) Customers wing** | | | |
| 10. | Office room | 30 | Documentation, records |
| 11. | Biofertilizer Storage room & Cold room | 50 | Racks etc. for bottle/cartoon storage |
| 12. | Selling room | 25 | Air conditioners.. |
| 13. | Passage, Toilet | 20 | Display etc. |
| **Sub Total** | | **125** | |
| **Total** | | **400** | |

**Table 2.14:** SWOT analysis for biofertilizer production

| **Strength** | **Weakness** |
|---|---|
| • Environmental friendly<br>• Positive impact on soil health<br>• Strong supply chain management<br>• Price value to the customers and qualitative products<br>• No adverse impact on human health | • New in the area<br>• Farmers are not habitual to use liquid biofertilizer<br>• No brand equity and image for new, small players and start ups. |
| **Opportunities** | **Threats** |
| • Less competitive market<br>• Easy availability of raw material<br>• New and emerging markets<br>• Futuristic move<br>• Export potential to neighboring countries<br>• Government subsidies | • Lack of skilled workers in the concerned area.<br>• Farmers demand for fast.results which is somewhat difficult with biofertilizers. |

## INTRODUCTION TO MICROBIAL BIOPESTICIDES

More than 25% crops get damaged due to pests and diseases annually leading to significant reduction in farmer's income and acts as a major threat to national food security. As a result, non-judicious use of pesticides by farmers to reduce their losses leads to environmental issues which could further lead to severe health hazards to humans, fauna and flora. There are severe issues regarding these pesticides, of which, their inherent nature of poor degradability in nature and their non-target effects on beneficial insects are the most threatening. To overcome these issues, natural products with their enhanced bio-degradability capacity is the need of hour. Biopesticides obtained from microbial sources could be an effective alternative of chemical pesticides (Table 2.15). Production of biopesticides on large scale is one of the demand driven areas of investment by the new agri-entrepreneurs to start up.

Biopesticides are living organisms and bio-control agents or their extracts, which are natural enemies of pests and thus kill the pests as predators, by infecting them or by producing toxins (Fig. 2.2).

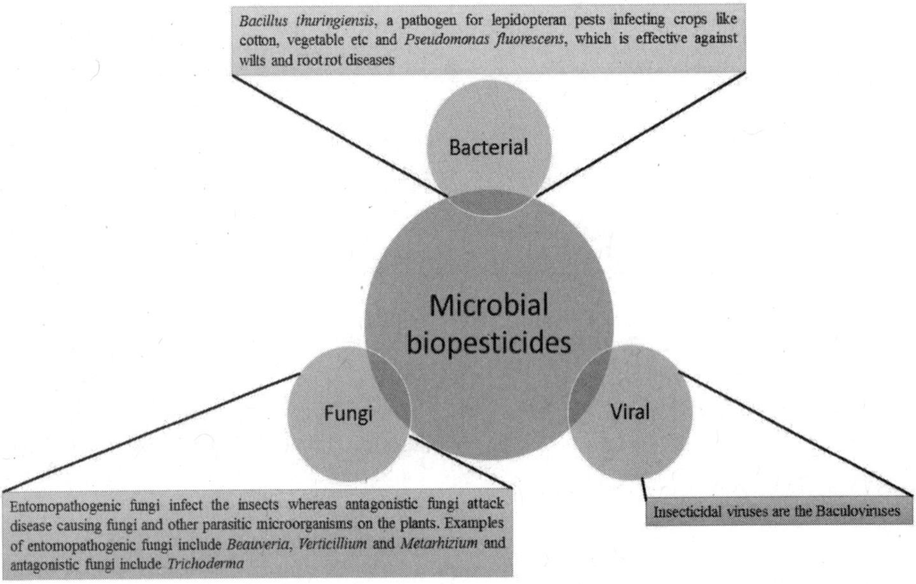


**Figure 2.2:** Types of biopesticides

**Table 2.15:** Comparison of biopesticide with chemical pesticide

| S.N. | Factors | Synthetic pesticides | Bio-pesticides |
|---|---|---|---|
| 1 | Cost effectiveness | Cheap, but increased spraying cost | Costlier, but reduced number of applications |
| 2 | Persistence and residual effect | High | Low |
| 3 | Knockdown effect | Immediate | Delayed |

| | | | |
|---|---|---|---|
| 4 | Handling and Bulkiness | Easy but danger and hazardous | Bulky: carrier based<br>Easy: liquid formulation |
| 5 | Pest resurgence | More | Less |
| 6 | Effect on beneficial flora | More harmful | Less harmful |
| 7 | Target specificity | Mostly broad spectrum | Mostly host specific |
| 8 | Nature of control | Curative | Preventive |
| 9 | Shelf life | More | Less |

*(Source: Agriculture Today, November, 2005)*

**Advantages of Biopesticides**

Biopesticides have the following advantages over chemical pesticides (Table 2.15):

1. They are eco-friendly as they degrade quickly and do not leave harmful residues.
2. They are more target specific affecting only the target pests and their close relatives and are non hazardous to other insects, fish, birds, mammals etc.
3. Development of resistance to biopesticides is less.
4. They are often effective in small quantity.
5. Biopesticides and bio-controls agents are relatively cheaper.
6. They are easy to apply and are suited for rural areas.
7. Following application of biopesticides there is no need for a waiting time before harvest
8. They are compatible with microbial biofertilizers, like nitrogen fixers and phosphate solubilizers.

The profile has been prepared for a unit of 100 TPA capacities on a single shift basis with the following product mix:

. *Trichoderma* : 50 TPA

. *Beauveria bassiana* : 30 TPA

. *Pseudomonas fluorescens* : 20 TPA

**Table 2.16:** Diseases controlled by biopesticides in various crops and method of application

| Biopesticide | Crop | Disease | Method of application |
|---|---|---|---|
| *Trichoderma viride* | **Cereals** | | |
| | Rice | Sheath blight | Foliar spray |
| | Wheat | Loose smut | Foliar spray |
| | Wheat | Root rot | Soil application |
| | **Pulses** | | |
| | Pea nut | Root rot | Seed treatment |
| | Black gram | Root rot | Seed treatment |
| | Chick pea | Wilt | Soil application |
| | Soyabean | Damping off | Soil application |
| | **Vegetable** | | |
| | Cauliflower | Damping off | Seed treatment<br>Soil application |
| | Tomato | Stemrot, damping off | Seedling dipping |
| | Brinjal | Collar rot | Seed dipping |
| | Peas | Damping off | Soil treatment |
| | Potato | Damping off | Tuber dipping |
| | Onion | Root rot | Soil application |
| *Pseudomonas fluorescens* | Rape seed | Root rot | Seed treatment or soil application |
| | Soya bean | Root rot | Seed treatment or soil application |
| | Cotton | Damping off | Seed treatment or soil application |
| | Cucumber | Damping off | Seed treatment or soil application |
| | Peas | Damping off | Seed treatment or soil application |
| | Chick pea | Wilt | Seed treatment or soil application |
| | Rice | Sheath | Foliar spray |
| | Pigeon pea | Blight | Seed treatment or soil application |
| | Tomato | Wilt | Seed treatment or soil application |
| | Linseed | Wilt | Seed treatment or soil application |
| | Banana | Wilt | Sucker dipping |
| *Beauveria bassiana* | Cotton | Pest control | Spray |
| | Pulses | Bollworm, whitefly cutworm | Spray |
| | Oil seeds | | Spray |
| | Vegetable | | Spray |
| | Tea | Flushworm, tea mosquito bug and shoot hole borer | Spray |
| | Rice | Rice hispa | Spray |
| | Coffee | Berry borer | Spray |

## Market

Biopesticides are the fastest growing segment of the organic agri inputs industry. The press release note by BCC research estimates global market of biopesticide to increase by CAGR of 12% i.e. from $2.1 billion in 2012 to $3.7 billion in 2017. However, chemical pesticides will have CAGR of 7% i.e. $ 44 billion in 2012 to $61.5 billion in 2017. These figures clearly suggest that the market for biopesticides is growing at a faster rate due to (i) increased awareness in users regarding the harmful effects of chemical pesticides and (ii) government policy for adopting environment friendly protectants.

In India, 16 biopesticides have been registered under the Insecticides Act, 1968. The biopesticide market represents 2.89% of the overall pesticide market in India and it is expected to increase annually at a rate of about 2.3% in the future (Rao *et al.*, 2015). Therefore, considerable increase in the use of biopesticides can be achieved with suitable awareness campaigns by the manufacturers as well as agricultural departments.

### Process Outline

Large scale of microbial biopesticides can be produced using fermentation technology in controlled conditions. Two types of fermentation system are available for microbial biopesticides (i) submerged fermentation for *Beauveria bassiana* and *Pseudomonas fluorescence* and (ii) solid state for *Trichoderma*. However, submerged fermentation is the preferred technology for all three microbes and hence will be discussed in coming sections. Some of the important considerations in the process of production are explained below:

1. **Fermentation process**: Submerged fermentation will be carried out utilizing the best nutritional medium for the optimum growth and production of uniform yield. The selection of raw materials should be cheap and easily available which should not decrease the overall yield and also have an ease of recovery. The overall cost of the products developed is dependent upon duration of fermentation, so one should take into consideration the fermentation variables which include pH, sterility, aeration, stirring etc.
2. **Formulation:** The inert material which facilitates easy transport, zero growth inhibitory as well as cost efficient should be used as a carrier for coating of living organisms or spores collected as an end fermentation product.
3. **Sterility:** The most important consideration during microbial product formulation is the sterility of the overall process. Sterility should be maintained and checked on a regular basis so to overcome reduction in yield at the end of the process.

## Project Details

### Production capacity

As mentioned earlier the profile has been prepared for a production capacity of 100 tons/annum on a single shift basis. The capacity is divided in the ratio of 30%, 40%, 40% among *Beauveria bassiana*, *Trichoderma viridae* and *Pseudomonas fluorescens* respectively.

### Location and site

The location of a biopesticide unit should not be in an area with extreme temperature conditions, otherwise it will increase the operating cost, resulting from controlled temperature conditions.

### Project cost

### A. Fixed Assets (Table 2.17)

**Table 2.17:** Fixed assets in biopesticides production

| S. No. | Description | Cost (Rs. in lakhs) |
|---|---|---|
| 1 | Land and site development | 5.00 |
| 2 | Building and civil works | 36.00 |
| 3 | Equipment | 39.80 |
| 4 | Miscellaneous fixed assets | 3.50 |
| Total | | **84.30** |

### I. Buildings and civil works (Table 2.18)

**Table 2.18:** Building and civil works for biopesticides plant

| S. No. | Type | Area (sft.) |
|---|---|---|
| 1 | Blending (2 x 500 sft) | 1000 |
| 2 | Packing and dispatch | 1000 |
| 3 | Administration | 1000 |
| 4 | Fermentation for entomo-pathogenic fungi | 1000 |
| 5 | Fermentation for antagonistic fungi | 1000 |
| 6 | Quality control laboratory | 1000 |
| 7 | Start culture preparation | 1200 |
| 8 | Inoculation | 600 |
| 9 | Mother culture and chemical storage | 1200 |
| Total | | **9000** |

### II. Equipment (Table 2.19)

**Table 2.19:** Equipments required for biopesticides plant

| S.N. | Name of the instrument/ equipment | Unit cost (Rs. in lakhs) | Quantity required | Total cost (Rs. in lakhs) |
|---|---|---|---|---|
| 1 | Autoclave | 4.50 | 1 | 4.50 |
| 2 | Packing machines 1 Kg | 0.10 | 1 | 0.10 |
| 3 | Packing machines 250 gm | 0.10 | 1 | 0.10 |
| 4 | Fermentor steady platform | 0.30 | 1 | 0.30 |
| 5 | Fermentor 200 litres | 15.00 | 1 | 15.00 |
| 6 | Air compressor with steam generator | 3.50 | 1 | 3.50 |
| 7 | Rotary shaker | 0.60 | 1 | 0.60 |
| 8 | Process water tank | 0.70 | 1 | 0.70 |
| 9 | BOD incubator | 1.50 | 1 | 1.50 |
| 10 | Hot air oven | 0.70 | 1 | 0.70 |
| 11 | Laminar air flow | 0.50 | 1 | 0.50 |
| 12 | Microscope stereo | 1.00 | 1 | 1.00 |
| 13 | Compound microscope | 1.75 | 1 | 1.75 |
| 14 | Refrigerator | 0.35 | 1 | 0.35 |
| 15 | Air conditioners | 0.30 | 4 | 1.20 |
| 16 | Electronic balance | 1.00 | 1 | 1.00 |
| 17 | Lab miscellaneous purchase | 7.00 | As per requirement | 7.00 |
| | | | Total | **39.80** |

**III. Miscellaneous fixed assets (computers, printers etc.): Rs. 3.50 lakhs**

A. Technology purchase – Rs. 10.00 lakhs

B. Working capital requirement

I. Raw material for growing biopesticide organisms – Rs. 33.00 lakhs

II. Manpower (Table 2.20)

**Table 2.20:** Manpower required for biopesticides plant

| Category | Total number of persons | Salary (Rs./month) | Cost (Rs./month) | Cost for salary (per annum) |
|---|---|---|---|---|
| **0.** General Manager | 1 | 50000 | 50000 | 600000 |
| **1.** Manager | 1 | 25000 | 25000 | 300000 |
| **2.** Scientist | 1 | 35000 | 35000 | 420000 |
| **3.** Supervisors | 2 | 15000 | 30000 | 360000 |

| | | | | | |
|---|---|---|---|---|---|
| **4.** | Plant operator | 1 | 12000 | 12000 | 144000 |
| **0.** | Technical assistants | 3 | 6000 | 18000 | 216000 |
| **0.** | Administration/Accounts | 1 | 5000 | 5000 | 60000 |
| **0.** | Marketing officers | 5 | 10000 | 50000 | 600000 |
| **0.** | Unskilled labour | 10 | 5000 | 50000 | 600000 |
| **Total** | | **25** | | **275000** | **3300000** |

III. **Utilities** (per month – Power 4000 units @ Rs. 6.5/unit) – Rs. 26000

IV. **Contingencies** (per month – marketing and maintenance expenses) – Rs. 50000

V. **Recurring expenses** (per month) –

| | |
|---|---|
| Raw material | Rs. 2.75 (in lakhs) |
| Manpower | Rs. 2.75 (in lakhs) |
| Utilities | Rs 0.26 (in lakhs) |
| Contingencies | Rs. 0.50 (in lakhs) |
| **Total** | **Rs. 6.26 (in lakhs)** |
| **Recurring expenses (per annum) -** | **Rs. 75.12 (in lakhs)** |

**C. Capital investment**

| | |
|---|---|
| Fixed assets | Rs. 84.30 (in lakhs) |
| Technology know how | Rs. 10.00 (in lakhs) |
| Working capital (3 months) | Rs. 18.78 (in lakhs) |
| Total | Rs. 113.08 (in lakhs) |

**D. Means of Finance**

| | |
|---|---|
| Debt | Rs. 67.83 (in lakhs) |
| Equity | Rs. 45.23 (in lakhs) |
| Total | Rs. 113.08 (in lakhs) |
| Debt Equity ratio | 3:2 |
| Interest rate | 16% |

**E. Financial analysis**

I. Cost of production

| | |
|---|---|
| Recurring cost (per annum) | Rs. 75.12 (in lakhs) |
| Depreciation (@10%) | Rs. 8.43 (in lakhs) |
| Interest (@16% per annum) | Rs. 10.85 (in lakhs) |
| Total | Rs. 94.40 (in lakhs) |

II. Turnover (Table 2.21)

**Table 2.21:** Turnover of biopesticides business

| Bio pesticide | Price (Rs./Kg.) | Quantity (Kg.) | Total (Rs. in lakhs) |
|---|---|---|---|
| *Trichoderma* | 195 | 40000 | 78.00 |
| *Beauveria* | 150 | 30000 | 45.00 |
| *Pseudomonas* | 125 | 40000 | 50.00 |
| | | **Total** | **173.00** |

III. **Profitability (turnover-cost of production)**

Net profit 78.60 (Rs. in lakhs)

**Table 2.22:** SWOT analysis for biopesticides

**Strength**

- Comparative advantage in biopesticides than chemical fertilizer
- Safety in formulating and applications
- Low cost of production.
- High quality
- Improved soil health
- Environmental sustainability
- Preserves traditional species without harming non-target organisms.

**Weakness**

- Productivity gaps
- Lack of established markets
- Poor quality management in production and processing
- Less incentives from Government
- Low R&D investments in identifying new strains and naturally occurring biopesticidal capability.
- Biopesticide market is buyers/ consumers driven market
- Disjointed producers, processors and traders

**Opportunities**

- Exponentially growing market.
- Growing purchasing power of consumers
- Growing health awareness
- Reduce heavy subsidies on food and fertilizers
- Earn high export earnings

**Threats**

- Lack of infrastructure facilities (like labs) and certification bodies
- Low awareness about biopesticides
- Most of the fields are contiguous and problem of contamination

## INTRODUCTION TO MICROPROPAGATION

Traditionally, a plant can be propagated by (a) asexual and (b) sexual means. Sexual propagation by seeds has a number of advantages of which easy transport and storage are the most prominent while asexual propagation by cuttings, rhizomes etc. helps in maintaining the purity of the plants. However, in sexual means, there is disadvantage of genetic variability and in asexual means (cutting, budding, grafting, layering etc.) limited number of plants can be produced.

Tissue culture or micropropagation is an asexual propagation technology which overcomes the above said problems in short duration.

**Definition**

Rapid clonal *in vitro* propagation of plants from cells, tissues or organs cultured aseptically on defined media contained in culture vessels maintained under controlled conditions of light and temperature.

The biggest advantage of micropropagation technology is acceptance by farmers all over the country owing to its uniform productivity, freedom from disease, vigorous growth and high yield. Higher yield is contributed by increase in the yield per plant as well as larger number of plants, which can be cultivated per unit area.

India has a wide range of climatic and soil conditions which enables the cultivation of an array of horticultural crops, floricultural plants, plantation crops, spices, medicinal and aromatic plants. Micropropagation technology has been used for multiplying various crops as shown in Table 2.23 below.

**Table 2.23:** Plants propagated through micropropagation

| Sr. No. | Plant group | Scientific name | Common name |
|---|---|---|---|
| 1 | Horticulture plants | *Musa acuminate* | Banana |
| | | *Fragaria* sp. | Strawberry |
| | | *Malus domestica* | Apple |
| | | *Punica granatum* | Pomegranate |
| | | *Saccharum officinarum* | Sugar cane |
| 2 | Spices | *Elettaria cardamomum* | Cardamom |
| | | *Vanilla planifolia* | Vanilla |
| | | *Zingiber officinale* | Ginger |
| | | *Curcuma longa* | Turmeric |
| 3 | Forestry plants | *Tectona grandis* | Teak |
| | | *Eucalyptus globules* | Eucalyptus |
| | | *Pinus roxburghii* | Pinus |
| | | *Bambusa vulgaris* | Bamboo |
| 4 | Medicinal plants | *Chlorophytum borivilianum* | Safed musli |
| | | *Stevia rebaudiana* | Stevia |
| 5 | Floriculture | *Orchidaceae* sp., | Orchids |
| | | *Dianthus caryophyllus* | Carnation |
| | | *Gerbera jamesonii* | Gerbera |
| | | *Anthurium andraeanum* | Anthurium |
| | | *Rosa* spp. | Roses |

## Market

Till date 100 commercial tissue culture production units have been recognized by the Department of Biotechnology (DBT), Govt. of India under the National Certification System for Tissue Culture Raised Plants (NCS-TCP). Their production capacity ranges between 1 million to 5 million and above plants per annum with an aggregate production capacity of more than 200 million plantlets per year. The protocols have either been developed in-house or transferred through the various research institutions and universities engaged in development of the protocols through support of the DBT. With increasing awareness about the advantages of tissue culture raised plants in improving yield and quality, their domestic consumption is also increasing optimistically. The annual market growth rate is expected at 20 to 25%.

## Product Technology

### Source of technology

Micropropagation technology is an expensive technology for any new initiator and therefore a number of laboratories, research institutions, micropropagation technology parks and the various tissue culture industries can be approached for transfer of technology. Two micropropagation technology parks have been set up at National Chemical Laboratory, Pune, Maharashtra and The Energy Resource Institute, Gualpahari, Gurgaon to develop commercially viable tissue culture protocols for various plant species. For example, date palm tissue culture protocol has been successfully developed by plant tissue culture laboratory, Anand Agricultural University, Anand; Gujarat is commercially available for licensing.

### Production Process

The process of plant tissue culture consists of the following steps as discussed below.

### Preparation of nutrient medium

The plant tissues are cultured on a semi-solid nutrient media prepared in double distilled water containing macroelements, microelements, amino acids, vitamins, iron source, carbon source like sucrose and phytohormones. The concentration of each of the above inorganic salts and organic supplements depends upon the protocol provided by the laboratory or developed by the in-house R&D unit. Depending upon the plant to be multiplied, different culture vessels are used like test tubes, small and large bottles etc. with required sufficient amount of nutrient media for optimum growth of cultures. The media is highly nutrient rich solution and therefore requires sterilization by autoclaving at 15 lb pressure for 15-16 minutes at 121°C.

### Establishment of aseptic culture

The starting material for the process is normally an actively growing shoot tip of auxiliary or terminal bud of a plant known as **explant**. The starting material should be selected with maximum care and precautions because through tissue culture technology all the plantlets produced will have the same genetic make-up as of mother plant. The explants will be excised from the mother plant and will be excised into smaller units followed by rinsing with tap water so as to remove the soil debris and particles etc. The excised clean explants will be further treated with different anti microbial agents to remove all the contamination present on the surface.

### Inoculation

The aseptic cultures are then inoculated/placed on nutrient medium under laminar air flow hood and then culture vessel lid will be closed tightly (Fig. 2.3)

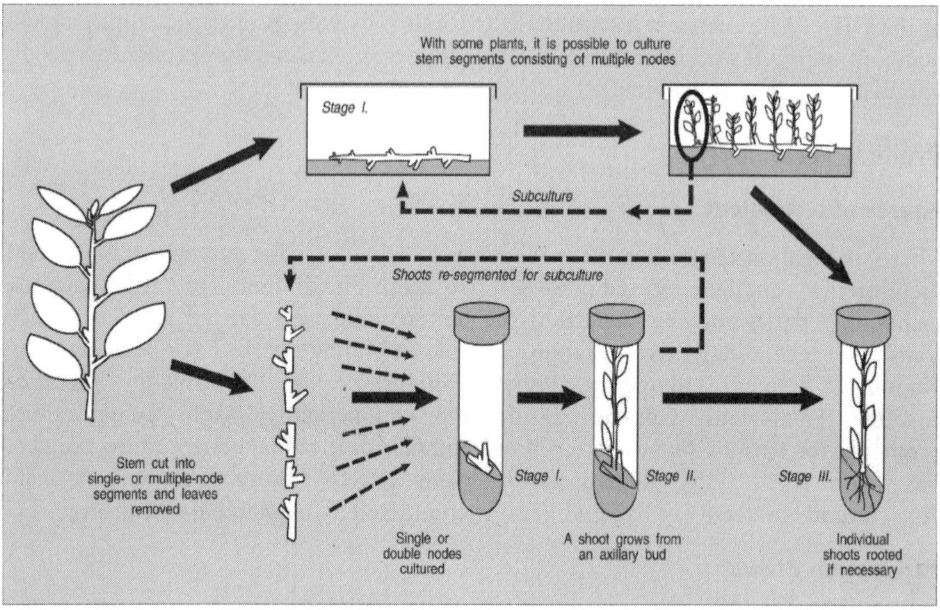


**Figure 2.3:** Single and multiple node culture
(*Source:* http://lib.znate.ru/docs/index-36441.html?page=3)

### Hardening plantlets

The plants are transferred to pots for acclimatization after they fully develop shoots and roots in the bottles. Hardening of *in vitro* raised plantlets is carried out under green house specially designed to maintain temperature and humidity so as to gradually harden the plantlets. The green house hardened plantlets are then transferred under poly/net house and fully hardened so as to supply to the farmers and growers. The duration of hardening under green and poly house is dependent upon the plant species.

## Project Details

### Production capacity

The below mentioned infrastructural requirement is based on unit with production capacity of 3 million plantlets per annum.

## Location

In order to facilitate marketing, the unit should be located near the area of cultivation of plants. Laboratory operations are independent of the location as these are done under controlled conditions in the culture room and growth rooms. The design of the greenhouse for hardening of the planting material will vary depending on the climatic conditions of the location of the unit. Alternatively only the hardening unit can be set up in the proximity of the market. Hardening unit can be an independent business unit.

## Project cost

### A. Fixed Assets

**Table 2.24:** Fixed assets required for a tissue culture lab

| Sr. No. | Head | Cost (Rs. in lakhs) |
|---|---|---|
| 1 | Land | 7.50 |
| 2 | Land development | 6.50 |
| 3 | Building | 40.00 |
| 4 | Utilities | 16.00 |
| 5 | Plant and machinery | 64.75 |
| 6 | Green and poly house | 40.00 |
| 7 | Miscellaneous fixed assets | 3.50 |
| **Total** | | **178.25** |

I. Land: The total area of land required for production of 3 million plantlets would be 4 acres with building and green house accounting for about 55% of the total land.

II. Land development (levelling, landscaping etc.) – 6.50 (Rs. in lakhs)

III. Building and civil works: The unit should have infrastructural facilities as mentioned below:

(a) Storage room for chemicals

(b) Media preparation room

(c) Sterilization and washing room

(d) Culture room

The total cost as estimated at Rs. 40.00 lakhs @ Rs. 400/sft

IV. Greenhouse: A greenhouse of area 7500 sq.ft. and a shade house of 80,000 sq.ft. will be constructed at a cost of Rs. 30.00 lakhs and 10.00 lakhs (total Rs. 40 lakhs) respectively. The greenhouse will be provided with heating equipment, fans and cooling systems

V. Equipments (Table 2.25)

**Table 2.25:** Equipments required for a tissue culture laboratory

| S.N. | Name of the instrument/ Equipments | Quantity | Unit price (Rs. in lakhs) | Total cost (Rs. in lakhs) |
|---|---|---|---|---|
| 1 | Autoclave | 2 | 4.75 | 9.50 |
| 2 | Laminar air flow | 10 | 0.75 | 7.50 |
| 3 | Magnetic stirrer | 2 | 0.15 | 0.30 |
| 4 | Hot air oven | 2 | 1.50 | 3.00 |
| 5 | Vacuum filtration device | 2 | 1.50 | 3.00 |
| 6 | Electronic weighing balance | 2 | 1.00 | 2.00 |
| 7 | Water purification system | 1 | 4.50 | 4.50 |
| 8 | Shakers | 2 | 1.30 | 2.60 |
| 9 | Drip irrigation/Fogger system | As per requirement | 5.00 | 5.00 |
| 10 | Air conditioners | 10 | 0.35 | 3.50 |
| 11 | Refrigerator | 2 | 0.80 | 1.60 |
| 12 | Stereomicroscope | 1 | 0.75 | 0.75 |
| 13 | Digital pH meter | 2 | 0.25 | 0.50 |
| 14 | Micro-oven | 2 | 0.25 | 0.50 |
| 15 | Shelves / Racks for cultures | As per requirement | -- | 10.00 |
| 16 | Laboratory glass-wares and culture vessels | As per requirement | -- | 8.00 |
| 17 | Green house miscellaneous items | As per requirement | -- | 2.50 |
| | | | **Total** | **55.75** |

VI. Miscellaneous fixed assets (computers, printers, etc.) – 3.50 (Rs. in lakhs)

**B. Working Capital requirement**

I. Raw material: The basic inputs for the production of micro-propagated plantlets include meristems of elite and disease free plants, ready to use culture medium and agar. Materials for media preparation include.

The basic raw materials for the project are:

i. Sucrose

ii. Agar agar

iii. Inorganic chemical compounds for supplementing micro and macro elements such as:

- Macroelements: $NH_4$ $NO_3$, $KNO_3$, $Cacl_2.$ $2H_2O$, $MgSO_4.7H_2O$, $KH_2PO_4$

- Microelements: $MnSO_4.4H_2O$, $H_3BO_3$, $ZnSO_4.7H_2O$, $Na_2Moo_4.2H_2O$, $CuSO_4.5H_2O$, $CoCl_2.6H_2O$ and KI.

iv. Glycine, Niacin, Pyridoxine.HCL, Thiamine. HCL, Myo-inositol for amino acids and vitamins supplements.

v. $FeSO_4.7H_2O$, $Na_2EDTA$ for iron source.

vi. Auxins, cytokinins and gibberllic acids for phytohormones.

The unit cost of raw material works out to Rs. 1.50 per plantlet. The total cost of raw material for 3 million tissue culture plants is estimated at Rs. 45 lakhs.

II. Manpower (Table 2.26)

**Table 2.26:** Manpower required for a tissue culture laboratory

| Name of post | Minimum number of posts required | Monthly salary (Rs.) | Total salary (per month) (Rs.) |
|---|---|---|---|
| Manager | 3 | 35000 | 105000 |
| Supervisor | 7 | 12000 | 84000 |
| Skilled | 22 | 6000 | 132000 |
| Unskilled | 11 | 5000 | 55000 |
| | | **Total** | **376000** |

**Total salary (per annum)** **Rs. 45.12 (in lakhs)**

III. Utilities
(Electricity 9000 units @ Rs. 6.5 per unit per month) – Rs. 0.58 (in lakhs)

IV. Contingencies
(Marketing, office expenses, repair etc. per month) – Rs. 0.75 (in lakhs)

**V. Recurring expenses**

| | |
|---|---|
| Raw material | Rs. 4.50 (in lakhs) |
| Manpower | Rs. 3.76 (in lakhs) |
| Utilities | Rs. 0.58 (in lakhs) |
| Contingencies | Rs. 0.75 (in lakhs) |
| Total | Rs. 9.59 (in lakhs) |
| **Recurring expenses (per annum)** | **Rs. 115.08 (in lakhs)** |

**C. Capital investment**

| | |
|---|---|
| Fixed assets | Rs. 178.25 (in lakhs) |
| Technology knowhow | Rs. 15.00 (in lakhs) |
| Working capital (3 months) | Rs. 28.77 (in lakhs) |
| **Total** | **Rs. 222.02 (in lakhs)** |

**D. Means of finance**

| | |
|---|---|
| Debt : | Rs. 133.21 (in lakhs) |
| Equity: | Rs. 88.80 (in lakhs) |
| Total: | Rs. 222.02 (in lakhs) |
| Debt: Equity | 3:2 |
| Interest | 16% |

**E. Financial analysis**

| | | |
|---|---|---|
| I. | Cost of production | |
| | Recurring cost (per annum) | Rs. 115.08 (in lakhs) |
| | Depreciation (@10%) | Rs. 17.82 (in lakhs) |
| | Interest (@16% per annum) | Rs. 21.31 (in lakhs) |
| | Total | Rs. 154.21 (in lakhs) |
| II. | Turnover | |
| | Average selling price | Rs. 12.00 per plant |
| | Total number of plants | Rs. 30 lakhs |
| | Total turnover | Rs. 360.00 (in lakhs) |
| III. | Profitability | |
| | Net profit | Rs. 205.79 (in lakhs) |

## Hardening Unit for Tissue Culture

A hardening unit for micropropagated plants can be set up by entrepreneurs at a lesser cost. The unit can also procure tissue culture plantlets from micropropagation laboratory and harden it in the facility and sell it to the farmers. The profile has been prepared for hardening facility for handling 3 lakh plantlets per annum.

### Project details

0. **Capacity:** 3 lakh plantlets / annum would require the following:
0. **Land:** 1 acre of land can be used for the development

### Project cost

**A. Fixed assets**

**Table 2.27:** Fixed assets required for a hardening unit

| Sr. No. | Expenditure Head | Cost (Rs. in lakhs) |
|---|---|---|
| 1 | Land and site development | 1.50 |
| 2 | Greenhouse | 10.00 |
| 3 | Electrical fittings | 0.75 |
| 4 | Furniture and fixtures | 0.60 |
| **Total** | | **12.85** |

**B. Working capital**

I. **Raw materials**

**Table 2.28:** Explants requirement

| Raw material | Rate (Rs. per plant) | Number of plants (lakhs) | Total (Rs. in lakhs) |
|---|---|---|---|
| Explant | 4.0 | 3 | **12.00** |

II. **Manpower**

**Table 2.29:** Manpower for a hardening unit

| Sr. No. | Category | No. of employees | Salary (Rs. per month) | Total salary (Rs. per month) |
|---|---|---|---|---|
| 1 | Sales and marketing | 1 | 15000 | 15000 |
| 2 | Greenhouse | 2 | 8500 | 17000 |
| 3 | Office and administration | 1 | 6500 | 6500 |
| | **Total** | **4** | | **38500** |

| | |
|---|---|
| **Total salary (per annum)** | **Rs. 4.62 lakhs** |
| III. **Utilities** | |
| (Power 500 units @ Rs. 3.0 per unit) | Rs. 3250 |
| IV. **Contingencies** | |
| (administrative, marketing etc. per month) | Rs. 10,000 |
| V. **Recurring expenses** | |
| Raw material | Rs. 1.00 (in lakhs) |
| Manpower | Rs. 0.39 (in lakhs) |
| Utilities | Rs. 0.03 (in lakhs) |
| Contingencies | Rs. 0.10 (in lakhs) |
| **Total** | **Rs. 1.52 (in lakhs)** |
| **Annual recurring expenses (per annum)** | **Rs. 18.24 lakhs** |

**C. Capital investment**

(Fixed assets + Working capital (3 months)) – **Rs. 17.41 lakhs**

**D. Means of finance**

| | |
|---|---|
| Debt | Rs. 10.44 (in lakhs) |
| Equity | Rs. 6.96 (in lakhs) |
| Total | Rs. 17.41 (in lakhs) |
| Debt : Equity | 3:2 |
| Rate of interest on loans | 16% |

**E. Financial analysis**

| | | |
|---|---|---|
| I. | **Cost of production (per annum)** | |
| | Recurring cost | Rs. 18.24 (in lakhs) |
| | Depreciation (@10%) | Rs. 1.28 (in lakhs) |
| | Interest (@16%) | Rs. 1.67 (in lakhs) |
| | (Debt: Equity = 3:2) | |
| II. | **Turn over** | |
| | Total plantlets | 3 lakhs |
| | Selling price | 10 per plant |
| | Total turnover | 30.00 lakhs |
| III. | **Profitability (turnover-cost of production)** | |
| | **Net profit** | **Rs. 11.76 lakhs** |

**Table 2.30:** SWOT analysis for micropropagation

| **Strength** | **Weakness** |
|---|---|
| • Independence regarding seasons and climate conditions.<br>• Lower need for agro-chemicals as plants are mostly disease free.<br>• Large scale of production from little starting material.<br>• Elimination of contaminants from the soil.<br>• Enables plant multiplication of species difficult to propagate through traditional methods. | • Complex acclimatization for the propagules and no adaptation to external climate.<br>• Possible presence of genetic variability.<br>• High initial production costs.<br>• Need for specialized labor.<br>• Serious contamination and vitrification is a plant reaction that makes them acquire a glass appearance, generating losses. |
| **Opportunities** | **Threats** |
| • Species more tolerant to climate change.<br>• In the future, plants with higher tolerance to acidic soils would be available.<br>• Stops current and continuous loss of plant diversity.<br>• Mass propagation of plants, in danger of extinction is possible.<br>• Permits determining an economic analysis of technologies of *in vitro* crops of other native forest species | • Loss of biodiversity as a consequence of displacement of traditional crops by a small number of genetically modified crops.<br>• More organization and competiveness of the agriculture sector in other countries that participate in the international market.<br>• In woody plants roots formed *in vitro* are not functional, which is why rooting adapted to the soil is necessary. |

## INTRODUCTION TO ENZYMES

Enzymes are proteins in nature and are natural biocatalysts produced by living cells. Enzymes are highly specific in their mode of action for e.g. cellulases degrade cellulose etc. due to their unique ability to recognize their substrate. This specificity enables enzymes for higher product formation with lesser byproducts synthesis which is the most important challenge in developing any industrial process. Different enzymes can be isolated and produced through protein engineering depending upon the process requirement like working in different temperature and pH (Table 2.31).

Due to their efficiency, specificity, the mild conditions in which they can work, their biodegradability, enzymes are well suited for a wide range of industrial applications.

**Table 2.31:** Widely used industrial enzymes

| S.N. | Industry | Enzymes | Application |
|---|---|---|---|
| 1 | Food processing | Amylase, protease | Production of grain based alcohol |
| | | Pectinase, glucose | Production of beer |
| | | Isomerase | Production of dextrose, maltose |
| | | Cellulase | Dextrins and starch derivatives |
| | | Lipase, papain, rennin | Fruit juice clarification, meat tenderization |
| 2 | Detergent | Protease, amylase | Improving washing efficiency |
| | | Lipase and cellulose | Stain removal |
| 3 | Textile | Amylase, cellulose | Desizing of starch, denim washing, territowel softening and biopolishing |
| 4 | Paper | Amylase, xylanase | Bleaching and sizing |
| 5 | Leather | Protease | Dehairing and bating |
| 6 | Animal Feed | Amylase, cellulose, hemicellulose and protease | Improvement in digestibility of feed for poultry and pork |

### Market

The biodegradability property with possibilities of protein engineering to produce on a large scale makes eznymes, most effective molecules to be commercially exploited. Enzymes play an important role in reducing the environmental pollution owing to their biodegradability nature and hence are regarded as environmentally safe biomolecules. The present enzyme market is more than 10,000 crores per year and which is increasing day by day with the introduction of novel enzymes that can replace synthetic hazardous chemicals. The leading manufacturers in India are:

0. Biocon Limited, Bangalore
0. Petrochemicals Industries Corporation (SPIC), Chennai
0. Advanced Biochemicals, Mumbai
0. Maps Biotech Limited, Ahmedabad
0. Textan Chemicals, Chennai

## Production Process

The production process (Fig. 2.4) consists of the following steps:

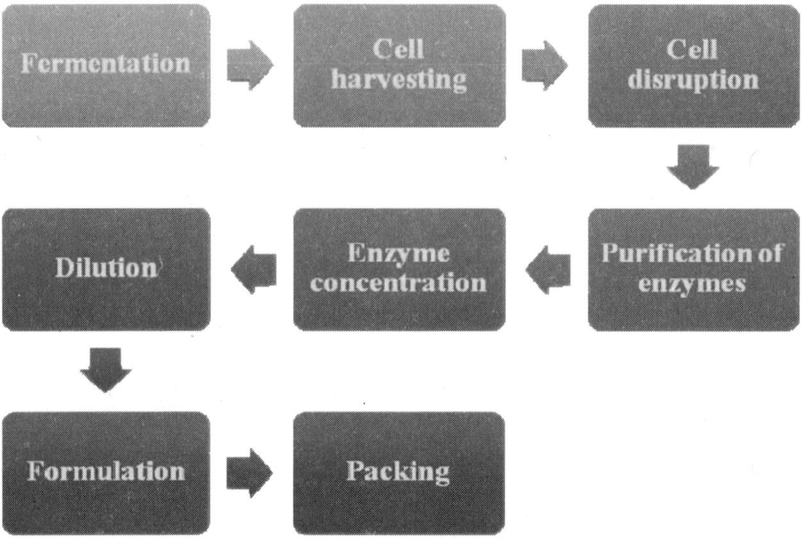


**Figure 2.4:** Commercial production of enzymes

## Project Details

### Production capacity

The present profile has been prepared for a production capacity of 1500 tones per annum (TPA) of enzyme bulk and formulations on a single shift basis.

### Area required

The land required is 2000 sqm. The cost of land has been assumed at Rs. 1000 per sqm. The total cost of land is estimated at Rs. 20 lakhs.

### Project cost

### A. Fixed assets

**Table 2.32:** Fixed assets in enzyme production

| Sr. No. | Description | Cost (Rs. in lakhs) |
|---|---|---|
| 1 | Land | 30.00 |
| 2 | Land development | 5.00 |
| 3 | Buildings | 50.00 |
| 4 | Equipment | 273.40 |
| 5 | Engineering cost | 159.00 |
| Total | | **517.40** |

I. **List and estimated cost of instruments / equipments**

**Table 2.33:** Estimated cost of instruments / equipments in enzyme production unit

| S.N. | Name of the instrument/ equipments | Quantity | Unit price (Rs. in lakhs) | Total cost (Rs. in lakhs) |
|---|---|---|---|---|
| 1 | Shaker (4 x 2 L) | 1 | 12.00 | 12.00 |
| 2 | Laminar air flow | 2 | 0.50 | 1.00 |
| 3 | Fermentors | 1 | 80.00 | 80.00 |
| 4 | Cell disruption system | 1 | 17.50 | 17.50 |
| 5 | Centrifuge | 1 | 12.50 | 12.50 |
| 6 | Seed fermentor | 1 | 15.00 | 15.00 |
| 7 | Autoclave | 1 | 4.50 | 4.50 |
| 8 | Dryheat sterilizer | 1 | 12.00 | 12.00 |
| 9 | Chromatography column | 1 | 4.50 | 4.50 |
| 10 | Ice maker | 1 | 5.00 | 5.00 |
| 11 | Colloid mill | 1 | 2.00 | 2.00 |
| 12 | Table top centrifuge | 1 | 0.50 | 0.50 |
| 13 | Concentrator | 1 | 1.00 | 1.00 |
| 14 | Peristaltic pump | 2 | 3.00 | 6.00 |
| 15 | Magnetic stirrer | 1 | 0.50 | 0.50 |
| 16 | Water bath | 1 | 0.30 | 0.30 |
| 17 | Membrane holder | 1 | 0.80 | 0.80 |
| 18 | pH meter | 2 | 0.20 | 0.40 |
| 19 | Conductivity meter | 2 | 0.65 | 1.30 |
| 20 | Spray dryer | 1 | 20.00 | 20.00 |
| 21 | Microfiltration | 1 | 20.00 | 20.00 |
| 22 | Ultrafiltration unit | 1 | 4.00 | 4.00 |
| 23 | Vacuum pump | 1 | 1.50 | 1.50 |
| 24 | Balances | 2 | 1.20 | 2.40 |
| 25 | Filter housings | 2 | 0.60 | 1.20 |
| 26 | Microscope (compound) | 1 | 1.50 | 1.50 |
| 27 | Cold room | 1 | 8.00 | 8.00 |
| 28 | Incubators | 1 | 2.50 | 2.50 |
| 29 | Movable mixing vessels | 2 | 6.00 | 12.00 |
| 30 | Mixing vessels | 2 | 10.00 | 20.00 |
| 31 | Ribbon blender | 1 | 3.50 | 3.50 |
| | | | Total | **273.40** |

II. **List and estimated cost of engineering services**

**Table 2.34:** Cost of engineering services required in enzyme production unit

| S.N. | Name of the instrument/ equipments | Quantity | Unit price (Rs. in lakhs) | Total cost (Rs. in lakhs) |
|---|---|---|---|---|
| 1 | D.G. set | 1 | 12.00 | 12.00 |
| 2 | Boiler | 1 | 10.00 | 10.00 |
| 3 | Compressed air | 1 | 6.00 | 6.00 |
| 4 | Water softening plant | 1 | 3.50 | 3.50 |
| 5 | RO plant | 1 | 10.00 | 10.00 |
| 6 | Distilled water plant | 1 | 15.00 | 15.00 |
| 7 | DW storage tank and SS circulation loop | 1 | 10.00 | 10.00 |
| 8 | SS kill tanks | 2 | 7.50 | 15.00 |
| 9 | Chiller | 1 | 4.00 | 4.00 |
| 10 | Incinerator | 1 | 9.50 | 9.50 |
| 11 | Dumping yard | -- | 3.50 | 3.50 |
| 12 | Gas bank with piping | -- | 6.50 | 6.50 |
| 13 | Work shop | -- | 2.50 | 2.50 |
| 14 | Safety and fire fighting | -- | 6.50 | 6.50 |
| 15 | Electricity connection | -- | 45.00 | 45.00 |
| | | | Total | **159.00** |

**B. Technology knowhow**

The cost of the technology varies depending on the number of enzymes identified for manufacturing. The cost will include design of the plant, the clone/ microorganism producing the enzyme and process detail.

**C. Working capital requirement**

I. **Manpower requirement**

**Table 2.35:** Manpower requirement in enzyme production unit

| Sr. No. | Category | No. of employees | Salary (per month) | Total salary (per month) |
|---|---|---|---|---|
| 1 | Manager | 2 | 50000 | 100000 |
| 2 | Supervisor | 3 | 30000 | 90000 |
| 3 | Administration | 2 | 20000 | 40000 |
| 4 | Unskilled labour | 10 | 6000 | 60000 |
| | **Total** | **17** | **106000** | **290000** |

| | |
|---|---|
| **Total salary** | **Rs. 14.64 lakhs per annum** |
| II. Raw materials (includes corn steep liquor, glucose etc.) | Rs. 85.00 (in lakhs) |
| III. Utilities (Power 2000 units @ Rs. 6.50 per month) | Rs. 0.33 (in lakhs) |
| IV. Contingencies (office expense, marketing etc. per month) | Rs. 0.85 (in lakhs) |
| V. Recurring expenses (I+II+III+IV per month) | Rs. 11.17 (in lakhs) |
| **Recurring expenses (V x 12 months)** | **Rs. 134.04 (in lakhs)** |

**D. Capital investment**

| | |
|---|---|
| Fixed assets | Rs. 517.40 (in lakhs) |
| Working capital (3 months) | Rs. 33.51 (in lakhs) |
| **Total** | **Rs. 550.91 (in lakhs)** |

**E. Means of finance**

| | |
|---|---|
| Equity | Rs. 8.71 (in lakhs) |
| Loan | Rs. 130.65 (in lakhs) |
| **Total** | **Rs 217.76 (in lakhs)** |
| Debt: Equity | 3:2 |
| Rate of interest | 16% |

**F. Finance analysis**

**I. Cost of production (per annum)**

| | |
|---|---|
| Capacity utilization (80%) | 1200 TPA |
| Total recurring cost | Rs. 134.04 (in lakhs) |
| Depreciation (@10%) | Rs. 51.74 (in lakhs) |
| Interest rate (@16%) | Rs. 20.90 (in lakhs) |
| **Total** | **Rs. 206.68 (in lakhs)** |

**II. Turnover**

**Table 2.36:** Turnover of enzyme production unit

| Sales (Tons) Breakup | Capacity production | Total amount (Rs./ton) | Sale price | Total (Rs. in lakhs) |
|---|---|---|---|---|
| Leather | 30% | 360 | 20000 | 72.00 |
| Textiles | 40% | 480 | 50000 | 240.00 |
| Feed | 15% | 180 | 75000 | 135.00 |
| Paper and pulp | 15% | 180 | 18000 | 32.40 |
| | | | Total | **479.40** |

**III. Profitability (turnover- cost of production)**

Net profit per annum Rs. 272.72 lakhs

**Table 2.37:** SWOT analysis for enzyme technology

| Strength | Weakness |
|---|---|
| • High quality<br>• Advanced downstream processing availability for maximum enzyme recovery.<br>• Safe bio-molecules<br>• Highly specific for substrate<br>• Least bi-products formation<br>• Environmental sustainability | • Productivity gaps<br>• Possibilities for poor quality management in production and processing<br>• Less incentives from Government<br>• Specialized enzyme market buyers/consumers driven market<br>• Disjointed producers, processors and traders<br>• Adulteration and poor quality raw materials availability.<br>• Intensive in nature and high R & D costs |
| **Opportunities** | **Threats** |
| • Big and growing market potential<br>• Growing purchasing power of consumers<br>• Growing health awareness<br>• High export earnings | • High cost of infrastructural establishment<br>• Costly and complex organic certification process<br>• Lack of model infrastructure facilities (like labs) and certification bodies<br>• Most of the fields are contiguous and problem of contamination |

*Chapter - 3*

# ENTREPRENEURSHIP OPPORTUNITIES IN FOOD PROCESSING SECTOR

**Samit Dutta**

*Associate Professor, College of Food Processing Technology & Bio Energy*
*Anand Agricultural University, Anand*

## INTRODUCTION

The food processing sector is a vital linkage and synergy between industry and agriculture and highly critical for the economy of any country. The Indian food value chain is on the verge of a great transformation from one characterized by high wastage, low processing and low global contribution to one that is more streamlined, more integrated and more significant in the global trade. Food sector in India has emerged as a high-growth and high-profit sector due to its immense potential for value addition, particularly within the food processing industry. Food processing involves any type of value addition to agricultural or horticultural produce and also includes processes such as grading, sorting, packaging, which enhance shelf life of food products. The food processing industry provides vital linkages and synergies between industry and agriculture.

In India, there are abundant opportunities for food processing entrepreneurs as food processing industry offers unique opportunities in production and export of processed food. The demand for various products is on the rise because of the increase in per capita income of the Indians which results in higher spending on value added foods. India is the world's second largest producer of food next to China, and has the potential of being the biggest in the food and agricultural sector. The food processing industry is one of the largest industries in India. It is ranked fifth in terms of production, consumption, export and expected growth. The existing problems of high levels of production and low levels of processing in this sector actually reveal the massive potential still rests to be exploited. The enormous growth potential of this sector can also be understood from the fact that food production in the country is expected to double in the next 10 years, while the consumption of value-added food products will also correspondingly grow.

The growth of food processing industry will bring immense benefits to the economy, raising agricultural yields, enhancing productivity, creating employment and raising life-standards of a large number of people across the country, especially those in rural areas. The liberalisation of the Indian economy and world trade and rising consumer prosperity has thrown up new opportunities for diversification in the food processing sector and opened new vistas for growth.

## A Rich Agricultural Resource Base

India has achieved significant growth in food production as a result of the green revolution. Over the years, agricultural production in India has consistently recorded higher output. The country has 127 agro-climatic zones (identified under National Agricultural Research Project), 195.25 million hectares of gross cropped area and 65.26 million hectare of net irrigated area (MiM, 2015). India ranks number one in the world in the production of milk, pulses, ginger, bananas, guavas, papayas and mangoes. Further, India ranks number two in the world in the production of rice, wheat, vegetable and horticulture products (MoFPI, 2015). Recent industry analysis reports of India Brand Equity Foundation (IBEF, 2015) mentions that India produced estimated 140 million tonnes of milk, 91,293,000 tonnes of fruits and 163,388,000 tonnes of vegetables during 2014-15.

India is the largest producer of buffalo meat (3.7 million tonnes in 2013-14) and the second largest producer of goat meat (0.16 million tonnes in 2013-14). It is world's second largest egg producer (65 billion) and third largest producer of broiler meat (3.8 million tonnes). With 9.6 million tonnes production in 2013-14, India is the second largest producer of fish in the world. India produces more than 200 million tonnes of different food grains every year. Total food grains production reached 251.12 million tonnes in 2013-2014. The total food production in India is likely to be doubled in the next 10 years. Despite the large production of food in India, food inflation and food security issues are major concerns for policy makers in the country as they affect the basic need for Indian citizens – to have sufficient, healthy and affordable food.

A nation-wide study on quantitative assessment of harvest and post-harvest losses for 46 agricultural produces in 106 randomly selected districts, carried out by The Central Institute of Post-Harvest Engineering and Technology (CIPHET), Ludhiana, has estimated harvest and post-harvest losses of major agricultural produces at national level was of the order of Rs. 44,143 crores per annum at 2009 wholesale prices. Depending upon the commodity, different stages such as harvesting, collection, threshing, grading/sorting, winnowing/cleaning, drying, packaging, transportation, and storage are considered for assessment of losses.

## Increasing Demand for Processed Food Products

On the other hand, India is undergoing transformation in its economy leading to changes in tastes and lifestyles, urbanization, and rising income levels. All of which are likely to have significant influences on food demand. With rapid increase in income and urbanization, food consumption in India has shown a pattern of change over the past three decades. From a diet primarily characterized by cereal, staple foods; mainly rice and wheat, to one that includes a larger share of milk and dairy products, fruit, eggs, fish, meat, as well as processed foods. As expected, there has been a continuous shift of food expenditure in favour of high value foods. The consumer expenditure survey, conducted by National Sample Survey Organization (NSSO) aims at generating estimates of household monthly per capita expenditure (MPCE) and its distribution, separately for the rural and urban sectors of the country, for States and Union Territories, and for different socio-economic groups. In June 2013, NSSO in its 68th round of survey has released the key indicators of household consumer expenditure in India, generated from the data collected during July 2011 - June 2012 (NSSO, 2013).

Consumption pattern trends during the last two decades indicate that since 1999-2000 in urban India and since 2011-12 in rural India non-food expenditure has surpassed the food expenditure, which is natural for any developing economy. Within food expenditure, the share of cereals has declined from 36.1% in 1972-73 and 25.6% in 1993-94 to 18.9% in 2011-12 in urban areas and from 55.6% in 1972-73 and 38.3% in 1993-94 to 24.7% in 2011-12 in rural areas. However, the share of milk and milk products within food expenditure has increased from 14.4% in 1972-73 and 17.9% in 1993-94 to 20.2% in 2011-12 in urban areas, making milk and milk product the largest food expenditure category. In rural areas, the share of milk and milk products within food expenditure has also increased from 10.0% in 1972-73 and 15.0% in 1993-94 to 18.7% in 2011-12.

Processed food products segment, including beverages, refreshment, cooked meal etc., has also grown significantly from 11.8% in 1972-73 and 13.2% in 1993-94 to 18.4% in 2011-12 in urban areas and from 3.3% in 1972-73 and 6.7% in 1993-94 to 11.9% in 2011-12 in rural areas, making it the third largest food expenditure category. So, in the demand side, processed and value added food products are going to be the key growth driver. Value addition of food products is expected to increase from the current 20% to 35% by the end of 2025.

## Indian Food Processing Sector – Current Status

Food processing industry in India is a sunrise sector, which has gained prominence in the recent years. According to the Ministry of Food Processing Industries (MoFPI), Government of India, the industries include items pertaining to these two processes viz. (i) **manufactured processes**-if any raw product of agriculture, animal husbandry or fisheries is transformed through a process (involving employees, power, machines or money) in such a way that its original physical properties undergo a change and if the transformed product is edible and has commercial value, then it comes within the domain of food processing industries and (ii) **other value-added processes-** hence, if there is significant value addition (increased shelf life, shelled and ready for consumption etc.) such produce also comes under food processing, even if it does not undergo manufacturing processes. Food processing sector can be categorized in different segments, such as, fruits and vegetables, meat, fish, poultry, dairy products, edible oils, beverages, grain mill products and other products like chocolates and cocoa products, soya-based products, mineral water, high protein foods etc. The Indian food processing industry accounts for 32 % of the country's total food market,

14% of manufacturing GDP, 13% of India's exports and 6% of total industrial investment (India in Business, 2016).

India's food processing sector is dominated by the unorganized sector accounting for 42% share, followed by small scale industries with 33% share and organized sector with 25% share (MoFPI, 2015). Though the market falls under the unorganized sector in the country, the organized sector has a larger share in the secondary processing segment than the primary one. Rice mills account for the largest share of processing units in the organized sector. The sizeable presence of small-scale industries points to the sector's role in employment generation.

## Export – Opportunities to be Tapped

The Indian food processing industry accounts for 13% of India's exports. India's export of processed food was Rs. 31,563.43 crores in 2014-15, which includes the share of products like mango pulp (Rs. 841.39 crores), dried and preserved vegetable (Rs. 847.11 crores), other processed fruit and vegetables (Rs. 2,569.93 crores), pulses (Rs. 1,209.51 crores), groundnuts (Rs. 4,675.35 crores), guar gum (Rs. 9,480.00 crores), jaggery & confectionary (Rs. 1,161.81 crores), cocoa products (Rs. 848.62 crores), cereal preparations (Rs. 3,038.79 crores), alcoholic and non-alcoholic beverages (Rs. 2,231.58 crores) and miscellaneous preparations (Rs. 2,437.77 crores) (APEDA, 2015).

Main export destinations for food products from India have been the Middle East and south-east Asia. Though India's contribution to world trade is miniscule, during the last five years, India's exports of processed food and related products increased at a CAGR of 23.3%. Growth has primarily resulted from greater exports to advanced economies; and more demand from emerging/developing economies as they experience strong growth. Growth in food product exports has been aided by significant improvements in product and packaging quality and greater private sector participation. India has a location advantage – it is geographically close to key export destinations (Middle East, south-east Asia). India's comparative advantage rests in its favourable climate, large agriculture sector and livestock base, long coastline, and inland water resources. India also has an edge in cost of production compared to its competitors in Asia and the developed world.

## Investment Scenario

According to the Department of Industrial Policies and Promotion (DIPP), the food processing sector in India has received around USD 6.55 billion worth of foreign investments during the period of April 2000 to September 2015 (India in Business, 2016). The Confederation of Indian Industry (CII) estimates that the food processing sector has the potential to attract as much as USD 33 billion of investment over the next 10 years and also generate employment of 9 million person-days. Gujarat Cooperative Milk Marketing Federation (GCMMF), popularly known as Amul, plans to invest Rs 5,000 crores to establish ten new processing plants as well as expand the current capacity to touch 32 million litres of milk per day capacity by 2020. American doughnut chain Dunkin' Donuts has tied up with local online grocery delivery platform Grofers for home-delivery of its packaged and freshly made products. Private equity firm India Value Fund Advisors (IVFA) plans to invest around USD 100-150 million in the food business in India over the next two years. Mahindra & Mahindra's agri-commodities division has announced its entry into pulses retailing under the brand **NuPro** with Tur dal being the first offering under the brand. ITC Limited plans to invest Rs 800 crores to set up a world-class food processing facility in Medak, a district located in Telangana. The company has also formulated plans to enter the dairy market.

## Policy Support for Further Growth

In order to promote food processing industries, increase level of processing and exploit the potential of domestic and international market for processed food products, Government of India has taken major initiatives in the last decade. 100% foreign direct investment (FDI) is permitted in the automatic route for most food products except for items reserved for micro and small enterprises. 100% FDI is permitted for alcoholic beverages, with the requirement of an industrial license. For pickles, mustard oil, groundnut oil and bread – items reserved for the micro small and medium sector, 24% foreign direct investment is allowed under the automatic route, with the requirement of prior approval from the Foreign Investment Promotion Board (FIPB) for FDI amounting to more than 24%.

Food processing was recognized as a priority sector in the new manufacturing policy of 2011. The government has introduced several schemes to provide financial assistance in the form of grants and subsidies for the setting up and modernization of food processing units, the creation of infrastructure, support for research and development and human resource development as well as other promotional measures to encourage growth within the processed food sector. The scheme for infrastructure development includes grant of subsidies for the setting up of mega food parks and integrated cold chains.

In order to promote faster establishment of food processing industries in the country, the government provides various tax and other incentives to business. In the budget 2015-16, a corpus of Rs. 2,000 crores was created under National Bank for Agriculture and Rural Development (NABARD) to provide cheaper credit to food processing industry. Excise duty on plant and machinery for packaging and processing has been brought down to 6% from 10%. The Government of India has planned to set up 42 mega food parks across the country in next three to four years.

## Conclusion

Indian food processing industry has seen significant growth and changes over the past few years, driven by changing trends in markets, consumer segments and regulations. Still there is a huge untapped market, at both domestic and international level, and these trends, such as changing demographics, growing population and rapid urbanization are expected to continue in the future and, therefore, will shape the demand for value added products and thus for food processing industry in India. The Government of India's focus towards food processing industry as a priority sector is expected to ensure policies to support investment in this sector and attract more FDI. India, having access to vast pool of natural resources and growing technical knowledge base, has strong comparative advantages over other nations in this industry. The food processing sector in India is clearly an attractive sector for investment and offers significant growth potential to entrepreneurs. Adoption of food safety and quality assurance mechanisms such as total quality management (TQM) including ISO 9000, ISO 22000, hazard analysis and critical control points (HACCP), good manufacturing practices (GMP) and good hygienic practices (GHP) by the food processing industry would enable adherence to stringent quality and hygiene norms and thereby protect consumer health, prepare the industry to face global competition, enhance product acceptance by overseas buyers and keep the industry technologically abreast of international best practices.

## References

1. APEDA (2015) "Processed Foods", Agricultural & Processed Food Products Export Development Authority, Ministry of Commerce & Industry, Govt. of India. Retrieved on December 31, 2015, http://apeda.gov.in/apedawebsite/six_head product/PFV_OPF.htm
2. IBEF (2015) "Food Processing" Industry Analysis Report of India Brand Equity Foundation, 15 August, 2015, Gurgaon, Haryana. Retrieved on December 22, 2015, www.ibef.org/download/Food-Processing-August-15.pdf
3. India in Business (2016) "Food Processing", Industrial Sector Report, India in Business, Investment and Technology Promotion Division, Ministry of External Affairs, Government of India. Retrieved on January 20, 2016, http://indiainbusiness.nic.in/newdesign/index.php?param=industryservices_landing/337/1
4. MiI (2015) "Food Processing Sector" Make in India initiative of Department of Industrial Policy & Promotion (DIPP), Ministry of Commerce, Government of India. Retrieved on December 31, 2015, http://www.makeinindia.com/sector/food-processing
5. MoFPI (2015) Annual Report 2014-15, Ministry of Food Processing Industries, Government of India, Panchsheel Bhawan, August Kranti Marg, New Delhi – 110049
6. NSSO (2013) Key indicators of household consumer expenditure in India, Round 68th, National Sample Survey Organization Ministry of Statistics and Programme Implementation, Government of India, New Delhi

*Chapter 4*

# ENTREPRENEURSHIP OPPORTUNTIES IN PROTECTED CULTIVATION

**Baljeet Singh**

*Doctoral Scholar, GH Patel Post Graduate Institute of Management, Sardar Patel University, VV Nagar, Gujarat*

## INTRODUCTION

Making vegetables available in the off season and flowers for export and domestic markets, represents a huge entrepreneurship in terms of protected cultivation business. People are ready to pay higher prices for vegetables in the off season. Similarly urban elites and upper-middle class with increased disposable income who somehow want to experience nature even in their busy city life and increased environment awareness has opened up new vistas for flower crops production under cover apart from immense export opportunity.

In terms of opportunity available, while demand supply gap would be studied with each crop, government support for protected cultivation including export oriented units has been quite encouraging in recent times and is being dealt here. The scheme is administered by National Horticulture Board (NHB) and is applicable to projects having area over 2500 sq. m. Pattern of assistance is credit linked back end subsidy @ 50% of the total project cost limited to Rs 56.00 lakh (as of Feb. 2015) per project as per admissible cost norms for green houses, shade net house, plastic tunnel, anti-bird/hail nets & cost of planting material etc.

The first and foremost requirement for an agri-entrepreneur to carry out protected cultivation to have a covered structure. Depending upon the location (agro-climate) and requirement (crop and purpose), it could be a polyhouse (crop cultivation), glasshouse (ornamental conservatory), lath house (nursery growing), shade net house (seedling hardening) etc. The common term used for these structures is greenhouse because of varying degrees of greenhouse effect being observed in these structures. Depending upon the affordability one can go for either low cost (manual operations), medium cost (semi-automatic or fan and pad type) or high cost (automatic and computerized). Brief description of these types is given below.

### Low-Cost Greenhouse/Polyhouse

The low-cost polyhouse is a zero-energy chamber made of polythene sheet of 700 gauge supported on bamboos with *sutli* and nails. Its size depends on the purpose of its utilization and availability of space. Like the greenhouse, it has one opening kept open for 1-2 hr during the day, especially in the morning to reduce the level of humidity inside. The temperature within polyhouse increases by 6-10°C more than outside. In UV-stabilized plastic film covered, pipe framed polyhouse; the day temperature is higher and night temperature lower than the outside. The solar radiation entering the polyhouse is 30-40% lower than that reaching the soil surface outside.

### Medium-Cost Greenhouse/Polyhouse

With a slightly higher cost, a quonset-shaped polyhouse (greenhouse) can be framed with GI pipe (class B) of 15 mm bore. This polyhouse has a single layer covering UV-stabilized polythene of 800 gauge. The exhaust fans are used for ventilation those are thermostatically controlled. Cooling pad is used for humidifying the air entering the polyhouse. It has a large exhaust fan fitted at one end and a cooling cell pad fitted at the other end. Cool air enters from the cell pad (which is continuously dripped with water) absorbs heat of the plants and gets expelled through exhaust fan as hot air. The polyhouse frame and glazing material have a life span of about 20 years and 2 years respectively.

### Hi-Tech Greenhouse/Polyhouse

This type of greenhouse consists of a sensor, a comparator and an operator. The temperature, humidity and light are automatically controlled. These are indicated through sensor or signal-receiver. Sensor measures the variables, compares the measurement to a standard value and finally recommends running the corresponding device. For example, temperature control system consists of temperature sensor, heating/cooling mechanism and thermostat-operated fan. Similarly, relative humidity is sensed through optical tagging devices. Boiler operation, irrigation and misting systems are operated under pressure sensing system. This modern structure is highly expensive, requiring qualified operators, maintenance, care and precautions.

## PROTECTED CULTIVATION OF ORNAMENTALS

Among the flower crops, rose, gladiolus, orchids, gerbera, are cultivated under cover mainly for export purposes. A part of production also caters to the local market mainly of continental hotels and high end institutional customers.

### Rose

Rose is considered as the queen of flowers and used for various purposes. It is used both as loose and as cut flower (with stalk). It is used for garland making, worship, decoration, gift etc. It is planted in edges, hedges, wall covers, rockeries, pots. Rose oil is used for massage; rose water is used as an eye drop/coolant. Gulkand is a mixture of sugar and rose petals, used a seasoning on bread. Pankhuri are dry rose petals. Gulrogan is hair oil having rose petals plus sesamum seeds. More than 1000 varieties of rose are grown in India. However, a very few of them are suited for export. As per plant form, there are three types- miniatures, hybrid teas and floribundas. Standards are again of three types- half, full and weeping standard.

**Demand supply gap**

Demand supply gap is huge for cut roses. Domestic supply mainly caters to continental hotels and high end events and important occasions, while the bigger chunk of cut roses grown under cover caters to the international market. Looking at the calendar of events, the maximum demand for cut roses falls in the months of October to February. However, for India, August is also included because of a number of festivals falling in the month. The valentine day requirement exports only for the year 2014 from India were pegged to the tune of 5-6 million stems compared to 4-4.5 million stems in the previous year (the Hindu Business Line 6$^{th}$ Feb. 2014). On an average the price for a single stem varies from Rs. 8-10 in the local market. Export market offers around Rs. 25 per stem on average (the Hindu 10$^{th}$ Feb. 2012). Going by these figures the export market for cut roses (for valentine day requirement only) stands at 12-15 crore. Total rose exports from India stood at Rs. 296 crores in 2010-11, Rs. 365 crores in 2011-12 and Rs. 423 crores in 2013-14 (Economic Times, 22 Jan. 2014). The export market for cut roses is growing steadily at a rate of 10-15% annually.

**Major markets**

Major international market is Aalsmeer in Netherlands while in India major market are rosebazaar.com (Bangalore) and Flower Trade Centre at Noida. Major export destinations of Indian cut roses are West Germany, Holland, U.K. and also US. Apart from Europe and USA, Middle East, Japan, Australia, Hongkong and Singapore import sizeable quantities of roses. Main competitors to India for export of roses are Columbia, Kenya, Equador and Israel.

**Varieties and planting material**

Main varieties dominating the rose scenario in India are First Red, Sonia, Vivaldi, Tineke, Melody, Darling, Only Love (Hybrid teas), Frisco, Mercedes, Jaguar, Kiss and Florence. Joy and Nikita are popular varieties of spray roses.

**Major players and market share**

IDL Floritech (Bangalore), Tanflora Infrastructure Park Limited (a joint venture between Tamil Nadu Industrial Development Corporation and MNA Associates promoting and supporting floriculture in the state), Vinayaka Agritech (Dharmapuri), Blooms and Greens (Bangalore), Soex Flora (Mumbai), Indo Bloom Limited (Bangalore), Elegant Flora (Mumbai), Hari Flora (Nasik), Dhatrak Group (Nasik),

Shrivardhan Biotech (Kolhapur), Nandi Floriculture, Devanhalli (Bangalore). Century International- 40 lakh stems annually (Noblesse, Grand Gala, Gold Strike) Suvarna Florex, Bangalore- 10 million stems.

**Cost of cultivation of rose under cover (per acre)**

3. Plant population: 27200
3. Cost per plant: Rs. 31
3. Planting material cost: Rs. 843200
3. Mortality (for) Rs. 42160
3. Total plants' cost: Rs. 885360
3. Fertilizers and plant protection: Rs. 30855
3. Labor: Rs. 14040
3. Bed preparation, pots, media: Rs. 400,000
3. Irrigation infrastructure including tube well and accessories: Rs. 300000
3. Farm equipment: Rs. 20,000
3. Store room/labor room: Rs. 20,000
3. Grading and packing room: Rs. 1,50,000
3. Cold storage including insulated panel: Rs. 10,00,000
3. Total: 2820255
3. Land 10%
3. Land development: 15%

**Note:**

Most of the planting material for export roses comes from foreign (Netherlands, France). Apart from the lump sum cost, our growers also have to pay royalty to these foreign breeders. The amount of royalty is around 1 pound per plant per variety (Market Insider 4th Aug. 2014).

**Operations and other requirements**

1. **Bed size:** 1-1.6 m wide, 30-40 m long and 15-20 cm height with 0.5-0.75 m space between two beds.
2. **Spacing:** 30-40 cm between rows, 14-18 cm between plants, 6-9 plants/m$^2$, 7 plants/m$^2$ is optimum, accommodate 70,000 plants/hectare.
3. **Media:** Both soil and soil-less substrates (rock-wool), pH 5.5-6.8, 30-40cm deep, well drained, porous, rich in organic content. Pasteurize with steam at 70-100°C for 30 minutes or use methyl bromide @ 10ml/cuft for 24-48 hr.
4. **Planting:** 6-18 month old budded plants are planted during May-June.
5. **Growth regulation:** Primary bending, secondary bending, gradual pruning, direct pruning, deshooting, disbudding etc.

6. **Manuring:** Depends on variety, type of medium used, growth stage, irrigation system etc.
   a. Nutrient composition of rose plant based on leaf analysis is 3% N, 0.2% P, 1.8% K, 1.0% Ca and 0.25% Mg.
   b. Nutrient requirement @ 1:0:2:1.2:0.3 NPK Mg.
   c. Fertigation requirement 170 ppm N, 34 ppm P, 160ppm K, 120 ppm Mg per every watering.
7. **Irrigation:** Yield starts 4-5 months after planting, harvest at tight bud stage for long distance, stem length vary from 40-90 cm. Flower yield range from 100-150 stems (hybrid teas), 200-225 (medium) and 250-350 (small and sprays) stems per $m^2$.

## Plant protection

1. **Pests:** Red spider mite, thrips, aphids, white flies, caterpillars.
2. **Management:** Scouting and instructing workers to mark infected spots (Table 4.1 and 4.2). Spray as soon as you see an infection (on the spot or whole greenhouse). Pest has to be in contact with chemical. Spray especially at bottom of the leaves. Use a lot of water (3000 till 4000 liter per hectare.) Spray with a high pressure, with a very fine mist. Spray again adult + nymph + larva + eggs at the same time. Spray same combination 2 till 3 times. Than it has to be replaced by other chemicals. By warm / hot weather the mites multiply very quick, so spray at least every 4-6 days in summer.

**Table 4.1:** Management of white fly in rose

| Trade name | Active ingredient | How to use |
|---|---|---|
| Chess | Pymethrozine | Spray |
| Ambush, Perthrine | Permethrin | Spray |
| Calypso | Thiacloprid | Spray * / ** |
| Applaud | Buprofezin | Spray * / ** |
| Admiral | Pyriproxifen | Spray * / ** |
| Confidor, Admire | Imidacloprid | Drench / trough drippers |
| Lannate | Methomyl | Spray / LVM |
| Dichlorvos , DDVP | Dichlorvos | LVM |
| Decis | Deltamethrin | LVM |
| Folimat | Omethoate | LVM |
| Sanmite | Pirydaben | LVM |

* When you spray Admiral, Applaud use double amount of water, and try to spray on the bottom side of the leaves.

** Save for most of the natural predators.

**Table 4.2:** Management of thrips in rose

| Trade name | Active ingredient | How to use |
|---|---|---|
| Orthene | Acephate | Spray |
| Success, Conserve | Spinosad | Spray |
| Vertimec, Avid | Abamectin | Spray |
| Regent, Violin | Fipronil | Spray |
| Lannate | Methomyl | Spray / LVM |
| Dichloorfos, DDVP | Dichloorfos | LVM |
| Folimat | Omethoate | LVM |

**Table 4.3:** Management of broad mite in rose

| Trade name | Active ingredient | How to use |
|---|---|---|
| Agramite, Floramite | Bifenazate | Spray |
| Vertimec, Avid | Abamectin | Spray |
| Kelthane | Dicofol | Spray |

**Table 4.4:** Management of two spotted mite / red spider mite in rose

| Trade name | Active ingredient | How to use |
|---|---|---|
| Agramitc, Floramite | Bifenazate | Spray |
| Nissorun, Calibre | Hexythiazox | Spray * |
| Apollo | Clofentezin | Spray * |
| Vertimec, Avid | Abamectin | Spray |
| Secure | Chlorfenapyr | Spray |
| Talstar, Kiros | Biphentrin | Spray |
| Pyranica | Tebufenpyrad | Spray |
| Folimat | Omethoate | LVM |

* Save for most of the natural predators.

**Table 4.5:** Management of aphids in rose

| Trade name | Active ingredient | How to use |
|---|---|---|
| Confidor, Admire | Imadacloprid | Spray |
| Orthene | Acephate | Spray |
| Pirimor | Pirimicarb | Spray |

**Table 4.6**: Management of sciarids / funges in rose

| Trade name | Active ingredient | How to use |
|---|---|---|
| Alsystan (Bayer) | Triflumuron | Drench |
| Suscon Green | Chlorpyrofos | Pellet |

1. **Diseases:** Powdery mildew, downy mildew, botrytis, black leaf spot, verticillium wilt.
2. **Management:** Spores of most of the fungi are always present (Table 4.7). If the circumstances are good for the fungi they will develop and the roots will start getting brown. If you drench the plants (+/- 100-150 ml water + chemical per plant) you have to make sure that the plants are thirsty. Before the drench/treatment stop the watering. After the drench/treatment you should not give water for half a day. You can give the drench/treatment early morning or instead of the last drip cycle.

**Table 4.7:** Management of white rust in rose

| Trade name | Active ingredient | How to use |
|---|---|---|
| Baycor | Bitertanol | Spray |
| Dithane | Mancozeb | Spray |

**Table 4.8:** Management of powdery mildew in rose

| Trade name | Active ingredient | How to use |
|---|---|---|
| Nimrod | Bupirimate | Spray * |
| Fungazil, Fungaflor | Imazalil | Spray * |
| Baycor | Bitertanol | Spray |
| Prosper | Spiroxamine | Spray |
| Rubigan | Fenarimol | Spray |

* You get the best result when you spray on a dark day with double amount of water.

Sulphur pots / burners are often used in the roses to reduce the change on powdery mildew, some natural predators are sensitive to sulphur.

## Sales and marketing

Around 25% of the cut rose is sourced by bigger firms/corporates from small farmers and exported. Rest of the exports route through big and progressive farmers, export firms having their own greenhouses. Price depends on the length of the stem. A cut rose having stem length 70 cm can fetch around Rs. 25 while that around one meter will fetch around Rs. 40.

## Returns

One rose plant can give out around 15 stems annually. Considering a plant population of 27,200 per acre, the total number of stems would be 4,08,000. The price of one stem being Rs. 25 in international market, it would fetch around Rs. 1 crore. Considering the cost of cultivation to be Rs. 30 lakhs/ acre, there would be a profit of Rs. 70 lakhs.

### Challenges

Among the pests, mites are a real havoc and also responsible for rejection of Indian roses in the international market. The rejections due to mites and other phytosanitary reasons could be as high as 25-30%. High input costs such as power and freight charges also add considerably to the costs. Besides, the rising labour costs are compounding the woes for the growers, who are seen pruning their output.

### Gladiolus

Gladiolus is also known as sword lily. It is a very popular bulbous ornamental plant. Its magnificent inflorescence with florets of dazzling colours, varying forms and sizes and long keeping quality makes it an attractive cut flower. Gladiolus spikes can be kept in the vases for 8-10 days depending on the variety and the ambient conditions prevailing in the room. Gladiolus is an essential component of most flower arrangements including bouquets.

### Demand supply gap

Protected cultivation of Gladious is done mainly for export because of better quality standards requirement.

### Major markets

United States, United Kingdom, Germany, Netherland and United Arab Emirates are the major importing countries.

### Major players

Thilak Agro Solutions (Bangalore), Spring Bloom (Bangalore), Flori Farm (Bangalore)

### Cost of cultivation (per acre)

1. No. of plants = 61000
2. Cost of each plant = Rs. 2.5
3. Cost of plants = Rs. 153000
4. Mortality = 7650
5. Total cost of plants = Rs. 160650
6. Plant protection and fertilizers = 20190
7. Labor = 15450
8. Irrigation infrastructure including tube-well and accessories = Rs. 50000
9. Farm equipment = Rs. 20000
10. Store room/ labor room = Rs. 20000
11. Grading, packing room = 20000

12. Total = 306290
13. Land = 10%
14. Land development = 15%
15. Total = 350000

## Operations and other practices

1. **Breaking corm dormancy:** Corms at 3 to 7°C for 3 months or treatment with Ethrel (1000 ppm) or $GA_3$ (100 ppm) or Thiourea (500 ppm) is adopted for breaking corm dormancy.
2. **Field preparation and planting**: Beds of size 6 x 2 m are prepared and corms are planted at a depth of 5 cm adopting a spacing of 40 x 25 cm (88,888 plants/ha) or 25 x 25 cm (1, 60,000 plants/hectare).
3. **Planting season:** October for plains and March-April for hills.
4. **Planting system:** Ridges and furrows system is adopted.
5. **Irrigation:** Irrigate at 7-10 day intervals in sandy soils and at less frequent intervals in heavy soils. Irrigation should be withheld at least 4-6 weeks before lifting of corms.
6. **Nutrition:** 120 kg N, 150 kg $P_2O_5$ and 150 kg $K_2O$ per hectare is recommended, of which 60 kg N and entire dose of $P_2O_5$ and $K_2O$ is applied as basal dose. The remaining N is given in two split doses, 30 and 60 days after planting.
7. **Fertilzers and manure:** Basal dose N 60 kg/hectare, P 150 kg/hectare, K150 kg/hectare.

Top dressing N alone is given @ 30 kg/ha during 4 leaf stage as foliar spray and 30 kg/ha during bud stage as soil application.

## After cultivation

After the corms have sprouted well, watering should be done, if necessary. When the shoots are about 20 cm high they are covered by heaping the soil up to a height of 10 to 15 cm. This enables the plants to grow erect despite high winds and rains and suppresses weed growth. Earthing up the soil is a must in case of light soils. In case where spikes grow longer or stems are not strong enough to bear the lodging or mild stroke of wind, they are supported with about 1.5 meters strong stakes. Strings instead of stakes may be used at the time of the appearance of the spikes. Strings are stretched between the stakes along the row to provide easy and adequate support.

## Plant protection

1. Before storage, corms are dipped in hot water at 40 - 45°C + fungicide (captan or thiram 2 g/lit) to control nematode and fungal diseases.
2. Thrips can be controlled by methyl demeton 25 EC @ 2 ml/lit. or dimethoate 30 EC @ 2 ml/lit.

3. Semilooper and *Helicoverpa* can be controlled by methyl demoton or monocrotophos @ 2 ml/lit.
   - **a. Leaf spot**: Spray carbendazim or mancozeb 2 g/lit to control leaf spot.
   - **b. Wilt:** Drenching of bavistin (0.2%) at fortnight intervals controls the wilt disease.
   - **c. Blight disease:** Blight disease can be controlled by spraying mancozeb @ 0.2 %.
   - **d. Storage rot:** Spraying of benomyl (0.2%) controls the storage rot.
   - **e. Fluoride injury:** Leaf scorch of gladioli is observed due to the presence of fluorine compounds in the atmosphere which got accumulated on the tips of leaves. The injury is associated with heavy application of super phosphate.
   - **f. Thrips:** Spray dimethoate 30 EC @ 2ml/l or fenitrothion 50 EC @ 2ml/l or malathion 50 EC @ 2ml/l.
   - **g. Semi-looper and Helicoverpa:** Methyl demeton 25 EC or monocrotophos @ 2ml/l or dimethoate 30 EC @ 2ml/l
   - **h. Harvesting of spikes:** Gladiolus takes 110-120 days to produce spikes. While harvesting, at least four basal leaves should be retained on the plant to ensure proper development of corms and cormels.
   - **i. Post-harvest treatment and grading**: Soak the stem in water to avoid wilting and lodging of stem and flower. Based on stem length and number of florets, the spikes are grouped into A, B, C, D grades.
   - **j. Sales and marketing**: In the domestic market gladioli are sold to florists and they also cater to demand of continental hotels and large institutional customers. Sales generally happen through agents. **Returns: 2.0 - 2.5** lakh flower stalks/hectare/crop. A single stem can fetch Rs. 1.5 to 2 in the local market and around 0.55 USD in EU market (except Netherlands), Japan (0.28 USD), China and Malaysia (0.44 **USD)** and 0.61 USD in the US market.
   - **k. Challenges:** Diseases, particularly virus are the main problem in gladiolus cultivation under cover. Once infested, it spreads very fast. So, it calls for daily inspection by trained personnel.

## Orchids

Orchids are accepted to be the world's most exotic and fascinating flowers, with their extraordinary variety of sizes, shapes, colours and markings. These features make them highly priced among the ornamentals. In India, they are mainly found in North Eastern hills, Kerala and Western Ghats and few in Kashmir. They grow in the humid tropical forests. Orchid growing has become a great hobby all over the world in addition to its commercial value for cut flower industry. The secret of successful cultivation of orchids is to provide them with conditions as identical as possible with the environment under which they were growing in the wild. Due to their peculiar nutritional habits, Orchids require special techniques of cultivation, if they are to grow healthy and put out a good crop of flowers.

## Climatic requirements

Orchids are broadly classified depending on the climatic zones in which they choose to grow; accordingly, three different types of orchids with different climatic zones have been recognized. North Bengal is said to be citadel of orchid production.

*Tropical zone (300-900m)*

This zone is characterized by dense, humid tropical forests with high amount of rainfall and humidity (90-100%). The temperature ranges from 22 to 27°C. This warm and humid belt is the home of epiphytic orchids, like *Aerides, Arundina, Ascocentrum, Dendrobium, Vanda teres etc.* (Kerala, Madras and other coastal places).

*Sub tropical zone (1800-3500m)*

This zone has mixed forests and receives less rain. Moss covered tree branches and rocks form a congenial habitat for orchids. The temperature ranges from 25-30°C and humidity almost 100% during monsoon. The most beautiful orchids occur in this zone. The epiphytic species of *Bulbophyllum, Coelogyne, Cymbidium, Dendrobium, Eria and Liparis* are found growing here. The terrestrial spps. like *Calanthe, Hebenaria, Phaius spps.* Some rare orchids like *C. gigantium, P. fairieanum* have also been reported from this zone (Western Ghats, South Karnataka).

*Temperate zone (1800-3500 M)*

The temperate zone remains covered with snow for 3-5 months a year. The temperature during summer varies from 10-14°C with Relative humidity between 80-100%. The trees of Rhododendron magnolia and Pinus are moss covered providing habitats for many epiphytic orchids like *Aerides maculosum, Arachnis clarkei, Bulbophyllum* spps., *Paphiopedilum* spps., *Vanda coerulea*, and over and above the excellent *Cymbidium spps*. and hybrids.

*Alpine zone (3500-5000 M)*

This zone, situated in the Himalayan range, is snow covered for about 4-6 months and mostly some terrestrial orchids are found to grow at lower ridges. Some of the orchids recorded form this zone are *Bulbophyllum retusiusculu, Habenaria cummin-siana, Herminium longilobautm, Nervilla macroglossa, Pleione maculate, etc.*

## Demand supply gap

There is a huge demand supply gap in the orchid market. In fact, India also imports sizeable quantities of certain orchids from Thailand and Malaysia. There has to be a weekly production of at least 2,000 orchids to cater for the domestic market and 10,000 orchids for the international market.

## Major markets and players

Major markets of orchids are in Holland, Australia, Thailand, Darjeeling Gardens Private Limited, Sliguri (West Bengal) and Guwahati (Assam) is a major player in production of orchids.

## Cost of cultivation

1. No. of plants per acre = 40800
2. Cost per plant = Rs. 60
3. Cost of plants = Rs. 2448000
4. Mortality = Rs. 122400
5. Total cost of plants = Rs. 2570400
6. Fertilizers and plant protection = Rs. 86380
7. Labor = 15300
8. Support system, bench, lighting = Rs. 400000
9. Cost of cultivation (Planting material and inputs) = 3072080
10. Bed preparation/pots and media = 400000
11. Irrigation infrastructure including tubewells and accessories = 300000
12. Farm equipments and other accessories = 20,000
13. Store room/labor room = 20000
14. Grading/packing room = 150000
15. Cold storage including insulated panel erection etc = 100000
16. Total = 4562080
17. Land = 10%
18. Land development = 15%
19. Total = 7000000

## Operations and other practices

Mass propagation of orchids is possible now-a-days through tissue culture, but lab requires heavy investments in the terms of building, specialized machines, instruments and skilled manpower. Project details about tissue culture lab have been discussed in chapter Entrepreneurship Opportunities in Agri Biotech. Orchids are parasites which feed on other plants to get their nutrition meaning thereby they do not have a photosynthesis mechanism of their own. Orchids should be grown in suitable containers according to their habitat. If you grow plants indoors and some require additional heat, this can be most conveniently supplied by heat boards or propagating beds. Provided the root area is kept warm, heating the whole growing area is not required; quite often there is enough radiant heat for the remainder of the plant. Inside most houses, the humidity is lower than most orchids appreciate. This can be increased by growing the plants on a tray of gravel with water kept to a level just below the surface of the gravel. Plants in pots standing on the gravel will benefit from the rising humidity. In warmer regions, if you grow plants outside, an area below the outer branches of a tree will often be satisfactory. Place the pots on a small bench or bricks to allow free circulation of air, yet prevent insects entering the pots. Most of the terrestrial orchids can be grown in soil or pots but most of the erect and epiphytic ones can be hung in baskets down the roof. They can also be arranged on the wooden logs or green fern and tied to the tree trunks if you have them in your compound. Trees in the garden form a natural Orchidarium.

Sunlight has several photomorphogenetic influences upon plants such as inducing photosynthesis, phototropism and photoperiodism. Light is essential for plants in many ways. It helps the mature plant to flower. Bud initiation in plants is the result of the action of a hormone, the phytochrome, which requires for its production, the correct duration of light and dark periods. It provides the energy necessary for the manufacture of carbohydrates.

**Light,** like many other cultural factors, will vary from genus to genus under cultivation. The optimum which orchids generally require is between 3000-6000 foot candles. *Cypripedium* and *Phalenopsis* need only between 200-300 foot candles and hence are to be kept in comparatively more shaded regions of the orchid house. *Cymbidium* will grow under full sun. Therefore, it becomes necessary to shade some orchid greenhouses to ensure that the proper amount of light will be available for good plant growth.

An intimate knowledge of the light requirements of orchids has led to the development of the fascinating procedure called **light gardening** which is now quite popular among orchid growers. There are particular regions in the spectrum of sunlight which aids particular processes in the physiology of plants. Thus the blue and orange-red regions of the spectrum aid in photosynthesis, while absorption in the far-red regions of the spectrum stimulates flowering. Based on this knowledge, extra amount of illumination is now being supplied to orchids either to supplement sunlight or to replace it entirely, with the result that more vigorous growth and greater output of flowers are achieved. Light gardening has made it possible to grow orchids even in cellars where no sunlight penetrates.

The **temperature** regimes used will be governed by the genera grown. Even though various species of orchids vary in their individual requirements of optimum temperature, generally orchids thrive in a day temperature varying between 16 to 21°C and a night temperature of 13 to 16°C. If plants are grown outside the tolerance limits, poor growth or even death will result. Again it must be stressed that temperature, light and humidity work very much in conjugation with one another and unless a harmonious relationship is achieved, optimum growth is not possible in any case. A widely held misconception is that all tropical orchids need really high temperatures, this is not altogether true and to subject them to such treatment can be disastrous. Three basic temperature regimes enable the enthusiast to grow nearly all cultivable orchids which are explained in Table 4.9.

Temperature influences not only proper vegetative growth, but also production of flowers in quite a good number of species. It should be noted that for optimum growth and flowering, the night temperature should decidedly be lower than day temperature, preferably by a margin of about 10°F. *Cattleya labiata* does not flower when night temperature is higher than 21°C. Most *Cymbidium* species need a temperature of 13-16°C, during nights and plenty of sunlight during day time for proper flowering. It has been reported that the *Dendrobium crumentaum,* flower only in response to a sudden drop in temperature of 10°C. However, in all cases, temperature changes must not be too sudden and too often. Very low temperatures common at the foothills kills plant tissues just as burning does.

**Table 4.9:** Climatic classification of orchids

| | | | |
|---|---|---|---|
| Cool species | Summer: Day | 16-21°C | *Cymbidiums, Odontoglossums,* |
| | Night | 13°C | *Paphiopedilums,* |
| | Winter: Day | 13-16°C | *Zygopetalums* thrive well. |
| | Night | 10°C | |
| Intermediate species | Summer: Day | 18-24°C | *Cattleyas, Laelias,* |
| | Night | 16-18°C | *Oncidiums,* |
| | | | *Stanhopeas and Dendrobiums* |
| | Winter: Day | 16-21°C | thrive well. |
| | Night | 13-16°C | |
| Warm species | Summer: Day | 21-29°C | *Phalaenopsis,* |
| | Night | 18-21°C | *Phaphiopedilums* and evergreen |
| | Winter: Day | 21-29°C | *Dendrobiums* thrive well. |
| | Night | 18-21°C | |

**Humidity** is very important for the good growth of all plants especially tropical ones and particularly epiphytes, which is much more difficult to control than either heat or light. It has been proved that about 50% humidity is suitable for most orchids. As a rule, they do not thrive in places where the relative humidity is less than 50%. In wild, the majority of orchids flourish in regions of perpetual mist. Since they are not regular soil plants, with a proper root system to absorb and supply them with enough moisture from the soil, epiphytic orchids do best in an atmosphere as saturated with water vapour as possible. Excessive humidity is dangerous; it brings about susceptibility to certain diseases. Plant parts may be attacked by various bacteria and fungi. Excessive high relative humidity (above 70%) contributes to succulent growth which is more susceptible to infection than hard growth. The danger of infection during the high relative humidity maintained for the purpose of cooling the plants in the daytime is lessened by good air movement. Humidity levels vary greatly from area to area and from country to country and it would be unwise to generalize. The most popularly used device for creating humidity is the evaporative water coolers. Spraying the benches, walls and floor of the orchid house or misting with very fine nozzles are also effective in this respect.

In stagnant air of a greenhouse, it is wise to reduce humidity at night. This is in complete contrast to natural conditions where the moisture content of the air is usually much greater during darkness but where it is counteracted by air movements which are absent in green house. As a general rule, any form of watering, clamping down or spraying should not be performed in the late afternoon or evening. In orchid houses we should run oscillating fans continuously. During the warmer months, an evaporated cooler keeps the orchids house day time temperature around 10°C and increases further air movement.

Good quality water is very important requirement to grow orchids successfully. Acceptable water sources include de-ionized distilled water and rain water. Watering needs depend on the potting media and the growing conditions. Correct watering has a direct relationship between atmospheric humidity and compost mixture. Fundamentally lower the atmospheric humidity; the damper should be the compost, and the higher the humidity the lower the moisture content of the potting mixture. A potting media should never be wet but just evenly moist. Occasionally short term drying out is beneficial as this closely imitates natural conditions whereby epiphytes are subject to very dry spells. Although it is not advisable to water pots daily, they should all be examined frequently and if too dry, well watering or even submerging by total dipping is suggested.

Terrestrial orchids require more water than epiphytes. Some orchids have heavy thick leaves with large pseudo bulbs; these are storage devices for water, adapted by the plant to suit quite dry environments. These plants must be allowed to partially dry out in between watering to have a wet/dry cycle of five to seven days. Other varieties have thin, soft leaves, requiring more frequent watering. Some like Pahpiopedilums do not have any pseudo bulbs at all, these therefore must be kept moist but not at all times wet.

Generally speaking, watering twice or thrice a week would be best. When drawing up such a schedule the following factors should be kept in mind.

1. Actively growing plants need more water than resting plants.
2. Large plants need more water than smaller ones or seedlings.
3. Plants grown in larger pots retain moisture for longer periods than those in smaller pots and hence need less frequent watering.
4. Plants require more water on a sunny day than on a cloudy day.
5. Plants growing in full sun utilize water more than those in shade.
6. Osmunda and coconut fibre retain moisture longer than either tree fern or bark, hence watered less frequently.

Ideally, plants should remain highly moist and should never be dry. More plants are lost to over watering that is watering too often-than by under watering. For most orchids, especially those with pseudo bulbs, a good rule is: "*If in doubt, do not water.*"

## Sales and marketing

Indian orchids have great demand in the hospitality sector like hotels and resorts, weddings and corporate events.

## Returns

Rs 350 and Rs 1,000 for a pot and Rs 100 to Rs 200 for a stem (2015).

## Challenges

Supply of planting material which hampers the steady production, apart from pests (aphids, caterpillars, mites, scales, thrips) and diseases (black rot, botrytis, leaf spot).

## Gerbera

It is also known as **Transvaal Daisy**, gerbera is one of the most important cut flowers, successfully grown under different conditions in several areas of the world and meeting the requirements of various markets. This success is primarily due to the wide range in colour and shape of the flower. The gerbera plant was first contributed by Robert Jameson in South Africa.

In the past, all gerberas where grown from seeds and cuttings. This changed in the seventies, when techniques for multiplication with in-vitro became available. These methods allowed a different approach in breeding and selection of gerbera.

## Demand supply gap

In recent years, there has been increased demand from European Union for Indian gerberas. The demand from the UK is increasing and there is substantial possibility of growth in floriculture. Potential market of the UK is not fully tapped. Interestingly, newer markets in Asia itself are also opening for Indian flowers like Singapore.

## Cost of cultivation

**Table 4.10:** Cost and returns of Gerbera production under protected cultivation

| S. No. | Particulars | Year | | | | | |
|---|---|---|---|---|---|---|---|
| | | I | II | III | IV | V | VI |
| 1 | Total cost | 6,86,700 | 1,35,000 | 1,45,000 | 2,57,000 | 1,55,000 | 1,55,000 |
| 2 | Flower yield/ poly house/year | 1,20,000 | 1,30,000 | 1,35,000 | 1,20,000 | 1,30,000 | 1,35,000 |
| 3 | Gross return @ Rs. 2.5/flower | 3,00,000 | 3,25,000 | 3,37,500 | 3,00,000 | 3,25,000 | 3,37,500 |
| 4 | **Net Returns** | **-3,66,700** | **1,90,000** | **1,92,500** | **43,000** | **1,70,000** | **1,82,500** |

**Note:** As the life span of the structure (excluding cladding material) is for about 10-15 years, the returns from 2nd year onwards will be substantially high.

## Operations and other practices

Gerbera's are mainly planted on a bench system. The space required between the walking paths is 75-80 cm and the recommended distance between rows is 75-80 cm.

The distance between the plants within the row should be 20 cm. The size of the distance between the plants is measured from heart to heart of the pot centre. A pot size of 3.5 / 4.5 litre and 18-20 cm deep is recommended.

## Soil

To achieve the best result in growth of the plants, soil preparation before planting is required. Most of the growers in Australia have worked with a bark mix in the past, but more and more growers change to a coco peat mix to fill there pots. As it is important to have a good water/air balance, for the coco peat a mix with 20-40% perlite is recommended. The pot must be filled 100% with the potting mix.

## Planting

Planting should be done in uniformly moistened potting mix. Place the plant into the potting mix so the top of the fertile pot is 2 cm higher than the potting mix. If planted too high, the plants might break at harvest. While planting too deep increases the risk of disease (rotting of the heart). Prevent root damage by carefully pressing the potting mix against the pot. Under conditions with high daytime temperatures in the greenhouse (>30°C), it is recommended to plant early in the morning or in the evening when the temperature are less extreme. To allow a good contact between pot and soil, it is recommended hand water very soon after planting.

## Water

A drip irrigation system is recommended; as each plant receives the same amount of water, and by supplying the water directly on the potting mix, the plant itself does not become wet (so preventing diseases). The pipes of the system are placed on the ground between the two rows, to prevent the dripper line becoming empty, and the water temperature in the dripper line is out the reach of direct sunlight.

Start with irrigation about one or two days before planting, to make the potting mix already wet. This will help the gerbera plant to make a better start. Start one hour after sunrise when the plants are just planted in the pots. In the first 1-2 months you must be careful that the potting mix does not get too wet, in this period you could stop 5-6 hours before sunset. When the plants are full grown you could stop 3-4 hours before sunset,

Use the drip irrigation 2-10 times a day. This depends on the size of the plant and the time of the year. There could be a difference, between a cloudy and sunny day of 40-50% in the usage of water with the gerbera plants. The drain must be between the 30-40% of the total water gift, please note that there could be a difference between the structures of the soil, these means that every type of soil requires a different amount of water.

Per dripper a minimum of 60cc and maximum of 100cc should be given per irrigation. When the plants are older a minimum of 80cc per irrigation should be given depending on the season. Check regularly if the moisture of the column just below the drip is the same as at the base of the column. If the top soil is wetter than the soil at the base, increase the water quantity per supply. On the other hand, if the situation is reverse (top soil drier than base), reduce the water gift. A capacity of 2 litres per hour is preferred as the chance of congestion is smaller.

By using a drip system, a wet (water) column is created through which the roots grow. Place the drippers by planting in the fertile pot, after 2-3 weeks when the roots are growing out of the fertile pot in to the potting soil replace them approximately 5 cm from the fertile pot.

## Fertilizers

A feeding unit with an A + B tank is preferred to give the exact quantity of nutrients to the plants. The EC and the pH are measured and directly corrected. The set points of the pH and EC put in the computer are given to the plants. Tables 4.11, 4.12, and 4.13 shows fertilizer feeding to gerbera plants.

**Table 4.11:** Fertilizer feeding to gerbera plants

**Tank: A = 1000 Litre**

| | | | | |
|---|---|---|---|---|
| Calcium Nitrate | CaNo3 | 26%CaO + 15.5 N | 75 | Kg |
| Ammonium Nitrate | NH4NO3 | 9%NO3 + 9% NH4 | 0-3 * | Kg |
| Iron Chelate | Fe | Edta 13% | 2 | Kg |
| | | or Eddha ** 6 % | 4 | Kg |

**Table 4.12:** Fertilizer feeding to gerbera plants

**Tank: B = 1000 litre**

| | | | | |
|---|---|---|---|---|
| Mono Potassium Phosphate | H2PO4 | 52% p2O5 + 34% K2O | 17 | Kg |
| Potassium Nitrate | KNO3 | 13% N + 46% K2O | 35 | Kg |
| Potassium Sulphate | K2SO4 | 52% K2O + 16% MgO | 2.5 | Kg |
| Epson Salt (Magnan Sulphate) | MgSO4 | 49%MgSO4 = 16% Mg0 | 25 | Kg |
| Nitric Acid | HNO3 | 38% | 0-10 *** | Litre |

**Table 4.13:** Fertilizer feeding to gerbera plants

**Trace elements (C )**

| | | | | |
|---|---|---|---|---|
| Manganese Sulphate | Mn | 32% Mn | 70 | Gram |
| Borax | B | 11% B | 250 | Gram |
| Zinc Sulphate | Zn | 23% Zn | 125 | Gram |
| Copper Sulphate | Cu | 25% Cu | 25 | Gram |
| Sodium Molybdate | Mo | 40% Mo | 25 | Gram |

* This depends of the PH of your drain water, if drain water PH below 5.5 no Ammonium Nitrate in the solution.

** When the PH from the drain water is above the 6.5, you have to use EDDHA iron.

*** This depends on the PH of the type of water you use e.g., rain, dam, bore water.

## Plant protection

Aphids, thrips, mites are the major pests on gerbera. We advice to spray/LVM during the warmer period every week with the chemicals which are killing a wide range of insects. If a problem occurs with a type of insect; spray a separate time with a chemical which is special made to kill that type of insect.

## Diseases

**Table 4.14:** Pythium and Phytophthora management in gerbera

| Trade name | Active ingredient | How to use |
|---|---|---|
| Fongarid | Furalaxyl | Drench |
| Previcur N | Propamocarb | Drench |
| Aliette | Fosetyl-Aluminium | Drench |
| Agri-fos Supa 400 | Phosphonic Acid | Drench |

**Table 4.15:** Fusarium management in gerbera

| Trade name | Active ingredient | How to use |
|---|---|---|
| Benlate | Benomyl | Drench |
| Bavistin | Carbendazim | Drench |
| Topsin M | Thiofanaat-methyl | Drench |

**Table 4.16:** Sclerotinia management in gerbera

| Trade name | Active ingredient | How to use |
|---|---|---|
| Rovral | Iprodion | Spray in the heart of the plants. |
| Sumisclex | Procymidone | Spray in the heart of the plants. |

**Table 4. 17:** Rhizoctonia management in gerbera

| Trade name | Active ingredient | How to use |
|---|---|---|
| Rizoclex | Tolclofos-methyl | Spray in the heart of the plants. |

**Table 4.18:** Botrytis management in gerbera

| Trade name | Active ingredient | How to use |
|---|---|---|
| Rovral | Iprodion | Spray |
| Scala | Pyrimethanil | Spray |

## Major markets and players

Domestic markets consume gerbera on marriages, special occasions, other functions, get together, parties, live plant gifting etc. Rise n Shine Biotech based out of Pune is a major player in Gerbera cultivation under cover and supply. Kala Biotech, Pune is involved in tissue culture production of gerbera plantlets. Advent Exim, Kolkata generally caters to the South East Asian Markets.

## Sales and marketing

Farmers around the cities sell gerbera to florists directly. Flowers brokers/agents also are present in the market and procure from small farmers.

## Returns

Around 200 standards and 400 miniatures can be produced in year per square meter. A single gerbera stem fetches around Rs. 10-12 in the domestic market. In off peak season prices come down to 5-6/ stem. Worst case secnario is Rs. 1-2/stem. Margin is around 100%. Flower brokers / agents charge typically 10% commission on sales.

## Challenges

Huge investment required to start the poly house and scarcity of skilled labor are the major problems haunting the gerbera cultivation industry, apart from quality planting material.

## Carnation

Carnations are most popular cut flowers in the global trade next to roses. They are popular due to their attractive colors, shades and keeping quality. Native of Mediterranean region is a recent introduction to India.

## Demand supply gap

Demand supply gap in carnation is huge. In 2003, world demand for carnation was 35 million stems which was an increase of 21% of the previous year.

## Operations and other practices

1. **Propagation:** Terminal stem tip cuttings of 4-5 nodes (10-15cm long) from vigorously growing, perpetually flowering plant roots, within 21 days. Stores cuttings at $0^{o}C$ for 40-60 days known to root better.
2. **Media**: Plants highly susceptible for water logged conditions and compactness. Therefore, need well drained, porous, light sandy loamy soils rich in organic matter are ideal. Require pH o 6.5 – 7.00.
3. **Bed preparations**: 20-25 cm height, one meter width and convenient length beds are prepared and sterilized with methyl bromide.
4. **Netting:** Nylon nets of 10 x 10cm (1 No.), 12.5x12.5cm or 15x15cm (4-5 Nos.) square are placed on bed before taking up planting. These nets are raised after planting at 15cm level intervals as and when the plants grows to provide support to weak stem with the help of fixed frames on either ends of the beds.
5. **Spacing**: Depends on the type of pinching and cultivar different spacing is adopted. For standards 20x15cm (33 plants/$m^2$) and sprays 20x20cm (25 plants/$m^2$) is given.
6. **Planting**: August-September planting is found ideal and reduce calyx splitting incidence. Roots are fibrous and only shallow planting should be taken up and watered immediately after planting.
7. **Irrigation**: Overhead irrigation, mist and fogging system is ideal for first 30-35 days of planting for better establishment of plants. When flower buds are appearing surface irrigation through drip is best. Excess watering is harmful.
8. **Nutrition:** Small doses in frequent application are better than heavy dose in a single application. The nutrient requirement depends on the media, type of cultivar, stage of growth etc.
    a. General applications is in the ratio of 1.0:0.5:1.5 NPK.
    b. Quantity of nutrients required per $m^2$ per year is 250g N, 80g $P_2O_2$, 200g $K_2O$, 125g Ca and 400g Mg. Applied in 4 splits at fortnightly intervals.
    c. Nutrient based on growth stages;
        i. From planting until the end of pinching: 15.5g N, 6.5g P, 14.5g K, 10.25g Ca and 2.25g Mg per $m^2$ per month applied at weekly intervals.
        ii. From end of pinching until stars of harvesting: 18.75g N, 6g P, 22.5g K, 11g Ca and 2g Mg per $m^2$ per month applied at weekly intervals.
        iii. During flower harvesting: 19.5g N, 6g P, 28.25K, 12.5g Ca and 2g Mg per $m^2$ per month applied at weekly intervals.

## Growth regulations

Depending on the cultivar and purpose different pinching methods are followed.

1. **First pinch:** When plants are 21-25 days old, terminal portion of the main shoot leaving basal 4-6 pairs of leaves. This method of pinching is called single pinch which intern produces 4-6 laterals.
2. **Second pinch:** It is done when plants are 45-50 days old or 21-25 days after first pinch. Here 2-3 laterals developed from single pinch (out of 4-6 nos.) are again pinched leaving 2-4 pair of leaves is removed or pinched off. This type of pinch is called one and a half pinch. These pinched laterals again produce 2-4 shoots each.
3. **Double pinch:** Here all the laterals produced after single pinch are pinched off to produce large number of shoots.
4. **Disbudding:** All the auxiliary and lateral buds are removed within 7 days of their appearance to ensure the terminal bud to produce a better quality flowers in standard cultivars. Whereas in sprays the terminal bud of the main axis is removed to encourage lateral shoots.
5. **Pests:** Red spider mite, aphids, thrips, moths, nematodes.
6. **Diseases:** Fusarium wilt and stem rot, Rhizactonia stem rot, Alternaria leaf spot, rust, bacterial wilt and blight phytopthora, virus diseases.
7. **Disorder:** Calyx splitting.

## Cost of cultivation

1. No. of plants/ acre = 1,19,000
2. Cost per plant = Rs. 8.5
3. Cost of plants = Rs. 1011500
4. Mortality = Rs. 50575
5. Total cost of planting material = Rs. 1062075
6. Plant protection and fertilizers = Rs. 72150
7. Labor = Rs. 22500
8. Support system = Rs. 300000
9. Bed preparation, pots and media = Rs. 400000
10. Irrigation infrastructure including tubewell and accessories = Rs 300000
11. Farm equipment = Rs. 20,000
12. Store/ labor room = Rs. 20,000
13. Grading/packing = Rs. 1,50,000
14. Cold store unit = Rs. 1000000
15. Total = 2946725
16. Land = 10%
17. Land development = 15%

## Major markets

From India carnations are exported to United Kingdom, UAE, Kuwait, Bahrain, Germany, Argentina, Switzerland, Saudi Arabia, New Zealand etc.

### Sales and marketing

Carnations are sold to florists. Some whole sale markets exist in India. Major consumption is on weddings and special events. With the growing agri-tourism industry in India, concept of U-pick is also gaining ground. Internationally, the main trends for the coming years are an increasing attention to corporate social responsibility (CSR), IT and e-commerce in the cut flower value chain and further concentration of retailers. Preconditions for a good position on the market are sustainably cultivated and transported flowers. Although there is only limited economic growth in the major markets in North-western Europe, demand in the eastern European markets is expected to grow. The demand for niche products will remain.

### Returns

Plants start yielding 110-120 days after planting and continue for 1½ to 2 years. The stage of harvest of flowers depends on type of market. Flower yield of 150-300 cut flowers per ms can be obtained.

## PROTECTED CULTIVATION OF VEGETABLES

Major vegetables grown under cover are bell pepper, cherry tomato, and cucumber.

### Tomato

Tomato is by far the most widely cultivated vegetable under cover. Record yields can be as high as ten times that under open field cultivation. Cherry tomato is also considered highly suitable for cultivation in green houses.

### Demand supply gap

Tomato grown in green houses is of high quality, good size and color and blemish free so finds a good demand in international market or continental hotels. Domestic consumption and demand for tomato is growing due to increase in population. It is very important vegetable with substantial nutritional value. Major export destinations are the neighboring countries of India. In North and South India, May-July and Dec-Jan is considered to be the lean season of tomato production. During this period greenhouse cultivated tomato can fetch higher prices.

### Operations and other practices

Cultural operations differ to a wide degree in covered cultivation as compared to open field cultivation of tomato. The reason being more judicious use of space, faster spread of pests and diseases in protected cultivation and high quality of output desired and different varieties (indeterminate) suitable for covered cultivation.

1. **Bed preparation:** Bring soil to a fine tilth by loosening it well. Beds should be one meter wide, 0.15 m high. Half meter working space should be left in between the beds for walking around and carrying out cultural operations.

   50, 50, 50 N, $P_2O_5$, $K_2O$/ ha should be applied to the soil. Disinfect the soil with 4% formaldehyde and cover with a poly sheet for 3-4 days. Apply polythene mulch. Make two inch diameter wide holes in the much as a spacing of 60X45 cms.

2. **Transplanting:** 35-40 days old healthy seedlings are transplanted. Transplanting should be taken up during morning or evening hours to protect the plants from heat shock. Place the seedlings in the centre of the hole and firm up the soil around gently.

3. **Fertigation:** Water soluble fertilizers like 19:19:19 should be mixed with irrigation water @ 250, 250, 250 Kg NPK per hectare starting from third week spread over the cropping period.

4. **Training and Pruning:** Tomatoes in green house need regular training and pruning. Leave two main stems; remove dead, rotten, criss crossed branches. Remove older leaves, if the live load is too high.

5. **Harvesting:** It starts around $10^{th}$ to $12^{th}$ week and then is carried out every alternate day (heavy crop load) or every third day (for medium crop load), 18-20 harvests can be expected. Yield is 150-175 tones per hectare.

**Plant protection**

1. **Diseases**: For seedling wilt spray Bavisitn 1 g/l or Ridomil MZ 2g/l. Powdery mildew can be treated with application of Karathane 1 ml/l. For blights, use DM-45 or DZ-87 @ 2g/l. In case of buck eye rot, apply Ridomil MZ @ 2 g/l. For viruses manage the insect vector.

2. **Pests**: For protection from white fly (vector for Potato virus X and Leaf curl virus) cover the nursery with 40 mesh nylon net. In case of red spider mite attack, spray with Dicofol 2 ml/l. Thrips- Acephate @ 1.5 g/l or Imidacloprid 0.5 ml/l.

**Major markets**

Tomato (both fresh and processed including frozen) is exported from India to UAE, Bangladesh, Oman, Nepal, Qatar, Singapore, Saudi Arabia, Benin, Israel, Australia, USA, South Africa, Mozambique, Angola, Sri Lanka, Maldives etc.

**Sales and marketing**

Generally, harvests from bigger greenhouses could be a full truck load. For long distance markets harvest at mature green or turning stage. Domestically these could be sent to bigger mandis like Azadpur vegetable market in Delhi or other whole sale markets. It is important to have good contacts with traders in mandis. Ripe tomatoes, can go to processing companies for sauce, ketch up, puree etc) provided they meet the specifications.

### Returns

Prices in local market are highly variable and can vary from Rs. 2 per Kg to Rs. 10 per Kg. The trick is to program production at the time of lean season. Considering a price of Rs. 5 per Kg, one can expect revenue of Rs. 8.5 lakhs from one hectare of protected tomato cultivation.

### Challenges

High input costs and unavailability of skilled labor are the major challenges in protected cultivation of tomato.

## Capsicum

Capsicum is the second most important vegetable crop grown under cover after tomato.

### Demand supply gap

Coloured capsicums are in great demand in urban markets. The demand is mostly driven by hotel and catering industry.

### Operations and other practices

California Wonder and Yolo Wonder are the capsicum varieties suited for green house cultivation. In general, indeterminate hybrids are grown under cover. Popularly grown commercial hybrids in India include Indra, Yamuna (Green); Bombay, Triple star, Natasha, Inspiration, Pasarella (Red); Sunnyez, Swarna, Orobelle, Bachata (Yellow). Places with high rainfall, humidity are not suitable (foliar disease spread is fast) nor are high wind velocity areas (green house structure may be damaged). Soil pH should be between 6 and 7. Net house is low cost structure with construction cost around Rs. 200/sq m, but yields are lower. Poly house is high cost structure with construction cost around Rs. 600/sq m. Yields are 15-20% higher compared to net house. Growing period is 8-10 months.

### Nursery raising

The seedlings are raised in pro-trays of 98 cells or cavities. About 16,000 to 20,000 seedlings are required to plant one acre for which 160-200 gm of seeds is required. The pro-trays are filled with sterilized coco peat and seeds are sown, one seed per cell to a depth of ½ cm and covered with the same media. Stack trays one over the above covered with plastic sheets. Seeds germinate in about a week's time after sowing. The trays are shifted to net house/ polyhouse and lightly watered. After 15 days of sowing, Mono ammonium phosphate (12:61:0) (3g/L) and 22 days after sowing 19:19:19 (3g/L) solution has to be drenched. The seedlings in portrays are drenched with COC 3g/L before transplanting. The seedlings will be ready for transplanting in 30-35 days. Spray imidacloprid @ 0.2 ml/L and chlorothalonil @ 1 gm/L before transplanting of seedlings. Always add about 0.3 ml/L of wetting agent per liter of water with each spray of pesticide.

**Land preparation**

Plough land thoroughly and mix with 10-12 tonnes FYM/compost per acre. Make raised beds 15-20 cm height, one meter width, keeping half a meter walking/working space in between the beds. Drench with 4% formaldehyde and cover with polysheet, leave for 3-4 days. Allow enough time for formaldehyde fumes to exhaust and turn over the soil well.

**Nutrient management**

20:25:20 NPK in the form of 80 kg calcium ammonium nitrate, 125 kg super phosphate and 32 kg murate of potash or 40 kg sulphate of potash.

**Laying of drip and mulch**

Place one 16mm inline drip lateral at the center of the bed having emitting points at every 30 cm interval with discharge rate of 2 ltr/hr or 4 ltr/hr. Black polyethylene non-recycled mulch film of 30-100 micron thick, 1.2 m wide, is used to cover the planting beds. Holes of 5 cm diameter are made on the polyethylene film as per the recommended spacing (45cm x 30cm).

**Transplanting**

35 days old seedlings are transplanted in already water (to field capacity) beds. Seedlings should be placed in the centre of the hole in the mulch without causing any damage to the roots. Roots should be placed at a depth of 4-5 cms. After transplanting, seedlings are drenched with 3 g/L copper oxy chloride or 3 g/L captan or 2 g/L copper hydroxide solution to the base of seedlings at the rate of 25-30 ml per plant

**Training and pruning**

The main stem of plant is tied with four plastic twine to train along and tied to GI wire grid provided on the top of the plants. This is practiced after four weeks of transplanting. The new branches and plants are trained along the plastic twines. The pruning is done after 30 days of transplanting at an interval of 8 to 10 days, resulting in bigger fruits with better quality and high productivity.

**Irrigation**

Irrigation is given through drip @ around 3 liters of water per sq m

**Fertigation**

Starting 3$^{rd}$ week of transplanting, provide water soluble fertilizers throughout crop growth as 19:19:19- 4kg, potassium nitrate- 1.5 kg and calcium nitrate 1.5 Kg/acre in each fertigation.

**Table 4.19:** Plant protection in capsicum under cover

| S.N. | Pest | Symptom | Management |
|---|---|---|---|
| 1 | Thrips | Upward curling of leaves, suck sap leading to stunted growth | Pongamia oil (5-8 ml/L) or Neem seeds kernel extract (NSKE 4%). fipronil (1ml/L) or chloropyriphos (2 ml/L) or acephate (1.5g/L) or Imidacloprid (0.5ml/L). |
| 2 | Mites | Downward curling of leaves, suck sap, reduced plant growth | spray Pongamia oil (5-8 ml/L) or Pongamia / Neem soap (8-10 g/L) or dicofol (2ml/L) or wettable sulphur (2ml/L) or abamectin (0.5ml/Ll) or ecomite or propargite or chlorophenapyr (1ml/ L) or fenazaquin (1 ml/L). |
| 3 | Aphids | Nymphs and adults suck sap from leaves, affecting plant growth, leaf curling seen | Spray Pongamia / Neem soap (8-10 g/L) or imidacloprid (0.5ml/L) or thiomethoxam (0.5g/L) or dimethoate (2ml/L). |
| 4 | Fruit borer | The adults lay eggs on fruits, flowers and leaves in large number and the nymphs that 14 come out of eggs, feed on fruits and affect marketability of produce | Spray thiodicarb (1ml/L) or carbaryl (3g/L) or indoxcarb (1ml/L) or rynaxypyr (0.5ml/L) or chlorofenfur (1.5ml/L) or fipronil (1ml/L) |
| 5 | Nematodes | Node like structures in the roots, reduced fruit size | Apply carbofuran (furadan) granules @ 20kg /acre at the time of planting, follow crop rotation |
| 1 | Damping off | Bases of young leaves are affected just above the ground level causing wilting and death of seedlings | Drench carbendazim (1g/L) or metalaxyl MZ (2g/L) or copper oxychloride (3g/L) or captan (3g/L) |
| 2 | Powdery Mildew | Powder like material on the lower surface leading to a powdery growth covering the entire lower surface of leaf which leads to drying and dropping of leaves at later stages. Reduced leaf and fruit growth. | Spray Pongamia /Neem oil (7ml/L) + sulphur WDG-80 (2g/L) or wettable sulphur (2g/L) or hexaconazole (0.5ml/L) or myclobutanil (1g/L) or dinocap (1 ml/L) |
| 3 | Cercospora leaf spot | initially as tiny yellow spot on leaf surface leading to increased dark grey spots this spreads on entire leaf resulting in dropping of leaf. | Spray chlorothalonil (2.5g/L) or mancozeb (2.5g/L) or carbendazim (1g/L) |
| 4 | Phytho-pthera | appears during fruiting and flowering stage resulting in tiny oil like spot on leaf surface resulting in rottening and blackening of plants | Spray copper hydroxy chloride (3g/L) or bordeaux mixture (1%) or metalaxyl MZ (2g/L) or dimethomorph + mancozeb (1 g + 2.5g/L) or fosetyl aluminium (2g/L) or azoxystrobin (0.5ml/L). |

**Cost of cultivation (Per acre of green house)**

1. Land preparation and FYM = Rs. 7688
2. Seeds = Rs. 2400
3. Treatment = Rs. 600
4. Plant Protection = Rs. 16000
5. Fertilizers = Rs. 7200
6. Labor cost = 40000
7. Misc cost = 6000
8. Rental value of land = Rs. 12000
9. Interest on fixed capital = Rs. 1200
10. Total cost = 93088

**Major markets**

Capsicums are exported from India to UAE, Singapore, Malaysia, Korea, Vietnam, Dominican Republic etc.

**Sales and marketing**

Harvesting starts 55 days after transplanting (for green), 75 days (for yellow) and 90 days (for red). 3-4 lobed god quality fruits with 150 g+ weight are graded as A, while 2-3 lobed good quality fruits with less than 150 g weight are grades as B. Fruits are packed in corrugated fiber board boxes and stored/transported at 7-8 degrees with 90-95% RH

**Returns**

The traditionally grown green capsicum, depending upon variety and season, usually yields 20-40 tons per hectare in about 4-5 months. In greenhouse, the crop duration of green and colored capsicums is about 7 -10 months and yields about 80-100 t per hectare.

**Challenges**

Major problems are perishability of the produce, diseases (damping off, powdery mildew, cercospora leaf spot, die back) and pests (aphids, thrips, mites and fruit borer).

■■■■

*Chapter 5*

# ENTREPRENEURSHIP OPPORTUNITIES IN FLORICULTURE AND LANDSCAPING

**Baljeet Singh**

*Doctoral Scholar, GH Patel Post Graduate Institute of Management, Sardar Patel University, VV Nagar, Gujarat*

## INTRODUCTION

Floriculture is an age old farming activity in India having immense potential for generating gainful self-employment among small and marginal farmers. In the recent years, it has emerged as a profitable agri-business in India and worldwide as improved standards of living and growing consciousness among the citizens across the globe to live in environment friendly atmosphere has led to an increase in the demand of floriculture products in the developed as well as in the developing countries worldwide. The production and trade of floriculture has increased consistently over the last 10 years.

Floricultural products consist of a wide variety of different types of plants and plant materials. They are broadly classified into four category, namely, (i) bulbs, tubers, corms, etc., chicory plant (non-food) (ii) other live plants, roots, cuttings (iii) cut flowers, dried flowers for bouquets, etc and (iv) foliage etc except flowers for ornamental purposes

### Floriculture Scenario in India

Floriculture activity has evolved as a viable and profitable alternative with a potential to generate remunerative self employment among small and marginal farmers and earn the much needed foreign exchange in developing countries like India. India has an ancient heritage when it comes to floriculture. Commercial floriculture is of recent origin. A consistent increase in demand for cut and potted flowers has made floriculture as one of the important commercial trades in Indian Agriculture. Around 80% of floriculture area is occupied by states like Tamil nadu, Karnataka, Andhra Pradesh, West Bengal, Haryana, Uttar Pradesh and Delhi.

### Landscape Architect and Landscape Designer

A landscape architect functions in the same manner as a landscape designer. He or she should be good at drawing, have the ability to analyze the site, a good knowledge of plants, flowers, color, environment and requirement of inputs besides that of pests and diseases and their management. However, there is no law that restricts landscape designing to only professionals. Thus any student or graduate can take up this profession. You should have love, respect and appreciation for nature. You need to understand that to work with plants, you need to dirty your hands in soil *(Ref: Landscape Gardening by Narendra Banker Vol. 1).*

### Annual Flower Seed Production

Flower seed production of annual flower crops (summer, rainy and winter) is another area open for entrepreneurship. Although anyone can take up the profession, but knowledge of plant breeding principles would be an added advantage. The first and foremost requirement for starting flower seed production business is the availability of land. The second point to be kept in mind is that flower seed production cannot be carried out at every place; rather there are specific pockets which support flower seed production. These are mainly governed by the agro climate in the region. Some of the existing players in flower seed production are Indo American Hybrid Seeds (IAHS), Pro agro (Bayer), Sutton Seeds, Chandrama Seed Farm (Jammu), Fredi Surti Company (Kolkata).

### Ornamental Nursery Business(B)

There has been an increasing demand for horticultural crops more particularly fruit and ornamental ones in both urban and rural areas of India. With this, the demand for good quality planting materials has gone up and hence the nursery business has developed rapidly in the recent years in our country. Nursery product is no longer restricted to orchards or large parks and gardens. It has entered into high rise buildings, offices, factories, business houses, hospitals, hotels, backyards, roadsides in cities, roof tops, etc. for decoration purpose. Heavy demand is observed during festive seasons and seasons of fairs etc. Ornamental nursery business has, therefore, come up in a large scale in areas near city and towns.

#### Pre-requisites

The site selected for raising a nursery should preferably be located near marketing centres for the convenience of transportation of the products with minimum or no damage. The site should be convenient enough for transportation of input materials also. It is important to have or develop a perennial source of water inside the nursery. If need arises, wind breaks of tall plants like eucalyptus, aonla, seedling mango, etc. may be planted to provide necessary shade and protection.

The product choice will primarily depend on the market demand in nearby areas. For wider market coverage, the choice may be dependent on market studies in the desired areas. Varieties of various ornamental plants like shade loving foliage plants, flowering plants, creepers, plants suitable for parks, gardens and roadside plantations, offices, business houses, hospitals, residential buildings, hotel, resorts, hostels, schools, colleges, Government offices etc. may be propagated in the nursery. Planting materials such as seedlings of flowers, bulbs, corms, etc. may also be produced.

**Dedicated semi skilled staff** having knowledge of plant propagation using cutting, grafting, layering, budding and acquainted with pests and diseases of plants would be an asset. Mother plants need to be procured from reliable sources. Provision needs to be made for workshed (6x4.5 m), poly house (9x4 m), store cum office (6x4.5)

## Gardening classes

Gardening is a good hobby, particularly after a long day's hard work, it refreshes you. In the morning hours, garden work is a good workout and may be considered equivalent to mild exercise. Housewives can utilize their time gardening in open space not only in the backyard or front portion of the house but also in pot. In metro cities roof gardening is also coming up in a big way. It also brings people close to nature who otherwise do not find time out of their busy scheduled to visit gardens, parks, forests etc. Gardening classes can be good business in metros, mini metros and bigger towns. To start gardening classes business you need to have a patch of fertile land, preferably on the outskirts of city with one classroom set up having audio-visual aids like LCD projector, computer etc. You also need to procure common garden tools like rake, spade, pots, hanging baskets, ververi. Having knowledge of agro chemicals, manures and fertilizers would be an added advantage.

## Bonsai Business

Bonsai is a very miniature tree, given a house plant treatment. Bonsai making is an artistic skill and bonsai are the end results of expert care and manipulation of a natural plant. You need to understand that bonsai décor houses but they are outdoor plants (miniature trees) and should be kept outdoors. Most of them fall under the category of temperates and even undergo dormancy.

Proper training and pruning skills is the first and foremost proficiency required for growing attractive bonsai. Root pruning is an important technique to slow down growth. Trunk height should be around six times the thickness of the trunk. All secondary branches should be alternatively at left and right. The branches should be wired down to create the illusion of an old tree. Proper manuring and fertilization should be done. Clay loams are the suitable soils. Root pruning and repotting needs to be done before the plants become rot bound. Soil should be replaced at the time of repotting. Less but frequent watering is desirable with proper drainage. As in any other tree crop, bonsai also respond to apical and lateral pruning.

To start the bonsai business, you need an open space to house the bonsais. Using benches would facilitate working and economise on space. You either procure mother plants from tree nurseries or grow trees from seeds. The latter one takes a longer duration but is more economical.

**Market and customers**

Apart from the regular brick and mortar shop space market, you can also sell plants online through ebay or just through your website. The buyer can pick the plants from your home or safe deliveries can also be arranged. Customers could be urban elites. Apart from being used as decorative pieces in open home spaces, terraces, window sills or in garden lawns, bonsai are also emerging fast as item of gift. But still, the authors would advise the readers to go for bonsai business if you're really passionate about it. It should be something you enjoy doing. At present, there are only a few organized players in the bonsai market.

## Interior Decoration with House Plants

This is also called **interiorscaping** or **interior plantscaping**. This is a billion dollar industry worldwide particularly in developed nations, but in India it is however, in infancy stage. If you are a plant lover, and enjoy beauty, then this is a fit business for you. You'll feed, water, prune and doctor indoor plants on a weekly basis for residential and commercial customers - from small offices to major shopping malls. You'll also lease plants, caring for them along with your other green charges, replacing them at no extra cost if they become unhealthy, and provide seasonal plant décor from winter holiday poinsettias to spring tulips. Minimal requirements are soil auger, watering cans, a pair of plant-snippers, plant foods agrochemicals and a good vehicle.

The market is urban elite households, continental hotels and institutions, commercial buildings and office space etc.

## Dry Flowers Technology

Dry flowers technology was first pioneered in India by National Botanical Research Institute (NBRI) Lucknow. Dry flowers may be used for making greeting cards, bookmarks, wall hangings, paperweights, table mats, dry-flower arrangements and so on. It does not require huge investments and can be started on home scale. The dry flowers-based, value-enhanced products cost Rs 30 to Rs 40 a piece normally fetch substantially higher prices in the market — even Rs 100 or more a piece if sold at airports and tourist shopping centres. The global market for dry flowers is relatively small, though it is expanding rapidly. Indian varieties of marigold, involve press-drying flowers in a microwave oven for 120 seconds. Calendulas need to be press-dried in a microwave oven for 90 seconds. Certain varieties of roses need to be embedded in sand and dried in hot air oven at 40 to 45°C for 48 hours to keep their original hues intact. Freshly-harvested chrysanthemums can be dehydrated by embedding them in silica gel and drying in hot air oven at 45 degree celsius for 48 hours.

## Florist

Floristry is again a multi billion dollar industry. For being a florist, you need to be an early riser as you need to go to whole sale flower market at least every alternate day.

You should also know flower arrangement, making bouquets both customized and standard. If you cater to events, which are where the highest profits are, you'll be designing the layout, negotiating prices, and coordinating transport and set-up. A flair for hygiene and creativity will add significantly to your business. Online market for flowers and their deliveries is increasing day by day. You need a small space but with definite visibility and higher foot fall. Most of the stuff is put on display rather than storing in the back room. However, some degree of temperature control should be practiced to prevent the flowers from drying and wilting. A short course in floral arrangements will increase efficiency and effectiveness of the business. You will really need to work harder on special occasions like Valentine's Day, Friendship day, Diwali, Holi etc.

## Essential Oils Extraction

Rose oil and jasmine oil are main products in this category. Oil is mainly extracted from damask rose while in jasmine, it is *Jasminum grandiflorum.*

Extraction of jasmine oil is carried out through enfleurage by mixing jasmine flowers with seasamum and spreading them for 10-12 hours. This is however, an old technique. In the process, seasamum seeds will absorb the perfume. Seeds are distilled and vapours of jasmine are absorbed into sandalwood oil for manufacture of attars. So, in essence, jasmine oil is not directly used in making perfume but it is the vapours.

A new method is solvent extraction using petroleum ether which results into concrete which can be directly marketed. Alternatively, concrete can be further refined with 95% alcohol to jasmine absolute. Concrete is used to make perfume blends and handkerchief perfumes. Other uses of concrete are in toiletries, confectionery, chewing tobacco, food essences and pharma industry. The leading variety of jasmine for concrete extraction is Jathimalli grown widely in Salem district of Tamilnadu. Cultivation of Jathimalli variety of jasmine for direct flower sale is highly profitable with an IRR of 29% as found by NABARD. Further, a start- up entrepreneur on a small scale can use the TNAU fabricated small scale concrete extraction plant for value addition to get jasmine concrete. On an average 1000 kg flowers would yield one kg concrete. Technology for jasmine concrete extraction is also available with Central Institute of Medicinal and Aromatic Plants (CIPMAP), Lucknow and JNTU Oil and Technological Research Institute, Anantpur.

Some of the manufacturers of jasmine concrete are Asean Aromatics (Chennai), Encee Aromatics (Metupalayam), Indian Flavors and Frangrances (Coimbatore). Jasmine concrete has a lot of export potential also particularly in European countries. Price of jasmine concrete in India is around Rs. 20,000 per Kg. These days synthetic jasmine essence made using jasmone, jasmone ester and jasmolactone also comes to market. Total annual demand for jasmine concrete is around 60 tonnes and the industry is growing at a rate of 7-8% annually.

## Protected Cultivation of Flower Crop

### Lawn and turfgrass management

Lawn and turfgrass management business is in a very infancy stage in India but a full fledged industry since decades in US, Europe and many other developed countries. According to the Professional Landcare Network (PLANET), an international association serving lawn care professionals, exterior maintenance contractors, installation/design/building professionals, and interiorscapers, there are an estimated 10,000 individual lawn care service providers and approximately 12,000 landscapers in the United States. In India, the business is confined to continental hotels, large commercial establishments, some tourist sites and golf courses. To be a lawn and turf grass manager you need to physically fit, should know gardening basics and should have a knowledge of varieties, hybrids, pests, diseases, nutritional and physiological disorders of grasses. Some of the grasses suitable for lawn making are: Soft/fine textured- *Cynodon dactylon* (doob), *Sporobolus tremulus* (Chain grass), *Zoisia tenuifolia* (Korean/velvet grass), rough/course textured- *Stenotaphrum secundatum* (St. Augustine), *Pennisetum clandestinum* (Kikiyu Grass). Kindly note that a lawn to thrive well needs open and full sunshine, though sometimes temporary wilting of grass may be seen in hot summer sun.

*Lawn making methods:* Predominantly, there are four methods to develop a lawn. These are elaborated below:

*Seeding*

The most popular grass suitable for seeding is Doob grass (*Cynodon dactylon)*. It has the fast spreading mat forming habit, radially forms roots at the nodes, the foliage is dark green, narrow with parallel vines. A lawn from seed is thought of only when grass roots are not available. About 30 kg of seed is required for planting one hectare. The soil should be reduced to fine tilth and given a light rolling. The site should be divided into suitable small squares or rectangles, the seeds are mixed with double the quantity of finely sieved soil and should be rolled again and watered liberally with rose can. The seeds take four to five weeks for germination. Care should be taken not to flood the site. For the first few times, the grasses are cut with a scythe. Lawn mower may be used for easy maintenance and for its spreading.

*Turfing*

The turfs are nothing but pieces of earth with compact grasses on them. These turfs should be cut uniformly in squares from a place where the grass is short, compact and free from weeds. These turfs should be placed on the prepared ground site, side by side and beaten down flat with a turf beater. The cavities in between should be filled with fine soil. The entire turfed area should be rolled and watered liberally. This is the most expensive way of lawn making.

*Turf plastering*

The doob grass can be procured in large quantities free from weeds and chopped properly into small bits of 5-7 cm long. Two baskets of chopped grass pieces should be mixed well with one basket each of garden soil and fresh cow dung and a shovel full of wood ash with required quantity of water to form a thick pasty substance. This mixture is then spread uniformly on the surface of a previously wetted perfectly leveled ground to a thickness of at least 2.5cm and watering should be done with a rose can. The next day, ground should be rolled and the grass should be allowed to spread. The grass will shoot up in a fortnight. To start with, cut with a scythe and after three months, use the lawn mower.

*Dibbling roots*

This is the cheapest, but time consuming method. Small pieces of grass roots should be dibbled 10 – 15 cm apart in a leveled ground when it is wet after rain. The roots spread and grow underground in the course of six months making a fairly compact lawn by frequent mowing, rolling and watering.

**Lawn management/after care**

Some of the important points which should be kept in mind during lawn management are enumerated below:

1. Fertilizing the lawn thrice a year is adequate to maintain rich greenness. Application of urea or ammonium sulphate at the rate of 1 kg / 50 sq. m during February - March, June - July and October - November is quite beneficial. At times well decomposed compost at 10 kg / 10 sq. m area will be sufficient as top dressing.
2. Weeds should be removed as soon as they appear, otherwise they spread, and seed multiply and overpower the grass. Fill the gaps with grass roots and fine soil. In the absence of rain, watering is done regularly at weekly intervals.
3. A mower should not be employed until a firm green sword has been formed. The grass is first cut with sickle and the surface is then rolled. Heavy roller should be used frequently but not when the ground is either too wet or dry.
4. Mowing should be done at brief intervals and never allow to produce seed stalks. Avoid cutting the grass too short as this can damage the grass, inhibit a deeper root system from setting up and give rise to weeds. Different grasses have different heights at which they can grow best, so make sure you enquire about this from the vendor or your landscaping company. The ideal height of most grasses is 3 to 4 inches. Removing more than one-third of the grass leaf in a single cutting is not recommended. Mow only on dry grass and not when the grass is wet. Make sure you mower's blade is a new and sharp before starting a fresh moving session. To ensure smoothness make sure you change the mower's oil once or twice during mowing season.
5. Once in a year, the lawn before rain and top dress with rich mixture of decomposed manure and soil. This will accelerate the grass with new vigorous growth.
6. The best time to water the lawn would be during the early hours of morning. Watering during this time will allow the water to reach the roots without

evaporating. Mid afternoons may lead to water getting evaporated soon and watering at night times can give rise to the possibilities of diseases. While watering, care should be taken to spread the water homogeneously across the lawn without over flooding or missing certain areas/spots. If the lawn is placed on heavy slopes make sure that the water does not run-off. Several applications of water would be necessary for such surfaces to ensure adequate penetration.

**Floriculture Inputs Business**

With the emergence of floriculture and landscpaing industry, floriculture inputs industry is also fast developing. These inputs include annual flower crop seeds, pots, hand painted pots, hanging baskets, flower vases, vase solutions, sprayers, fertilizers, biofertilizers, nutraceuticals, plant protection chemicals, sphagnum moss sticks, vases, growth media, vase solutions, machines like secateurs, shovels, shade nets, polyfilms, plastic (seedling) trays etc. Floriculture inputs business is likely to do well in metros and mini metros. Again, to excel in this business you should have knowledge of cultivation aspects and package of practices of ornamental horticulture crops.

Floriculture inputs business offers a good margin. The profit potential can be imagined from the fact that a small plastic pot having 27 cm upper diameter can be priced anywhere between Rs. 140 to Rs. 1500 depending upon design and strength whose cost of production in no case is more than Rs. 100. Similar is the case of hanging baskets. Among the pesticides and fungicides, MNCs offer lesser margin as compared to local counter parts. Also, nutrient supplements have bigger room to earn higher margins. You need to be a direct distributor of some of the companies (5-6 is manageable) and retailer for others that means for these you will have to take your supplies form distributors who are directly linked to the company. It is better to keep retail business of more companies that direct distributorship. Unit margins are lesser in distributorship. It is better to have the distributorship of those items which you can sell in bulk or in higher volumes.

Considering advisory an input, it can be offered to larger floriculture farms, green houses, export oriented units, mainly on plant protection. Information on markets, flower shows, events can also be given. It is better to have a set of clientele on an ongoing basis who in the long run become loyal customers and would contribute to ensuring sustainability of your business.

Another advantage of input business is that many input companies offer volume sale based/ quantity based incentive schemes which can even be foreign trips. This not only offers tourist opportunities but exposure visits which are important from point of view of enhancing your technical knowledge in the trade and also to know international/best practices in floriculture business. Details of investment and returns are discussed in sub-chapter on agri retailing.

## Suggested Reading

1. http://www.entrepreneur.com/article/190462 (assessed on 15th Jan. 2016 at 12.37pm).
2. http://agritech.tnau.ac.in/horticulture/horti_Landscaping_lawn%20making.html (assessed on 15th Jan. 2016 at 12.41 pm).
3. http://www.entrepreneur.com/businessideas/hand-painted-pots (assessed on 19th Jan. 2016 at 10.09 pm).
4. http://www.gardeners.com/buy/outdoor-planters/flower-pots-planters.

*Chapter 6*

# ENTREPRENEURSHIP OPPORTUNITIES IN SUPPLY CHAIN MANAGEMENT OF AGRI-PRODUCE

**Samarendra Mohapatra**

*Professor, OUAT, Bhubaneswar, Odisha*

## INTRODUCTION

India's population has crossed 1.25 billion and nearly 65% is dependent on agriculture and allied sectors. Yet, the contribution from this sector to the country's gross domestic product (GDP) is only around 13.7%. India is the second-largest producer of agricultural produce such as fruits, vegetables, flowers and raw silk. We also produce massive quantities of wheat, rice, sugarcane, oil seeds and palm oil. But where is the advanced agro-processing, the value addition, the forward integration? One industry that showed true potential initially was the milk industry. The classic example is Amul, whose success was largely based on two fundamental tenets Mr. Verghese Kurien propagated - one was seeding the sense of ownership the farmers felt about the cooperative society and the other was building a strong brand in the form of Amul. India's milk output has reached an estimated 140.6 million tonnes in 2014. With this, India has come a long way from being a milk deficient country in the 1970s to becoming the world's largest milk producer and consumer today. The lessons a small entrepreneur can derive and implement out of this cannot only transform the way he looks at the business operations, but also the deep-rooted relationship and interdependence between various factors in an ecosystem.

Value chain analysis in an upgraded term is called **value chain mapping**. Agri-entrepreneurs can add value to preliminary agri-products through the combinations with other resources such as tools, manpower, knowledge, skills, and raw materials. As products passes through several stages of the value chain, the value of the product increases as illustrated in the Figure 6.1.

One needs to realize that the entrepreneur's stake in developing his ecosystem is as important as the ecosystem's stake in a small entrepreneur's development. There are many sub-sectors where there is a lot of untapped opportunity such as in poultry, fruits, cotton and horticulture. Take fruits, and see what european brands have done by means of value addition. Fruits like strawberry, peach, kiwi, etc., can fetch a syrup manufacturer 30 times more value per kg than it would to the farmer who grows them. The questions arises why could we have not done this in India? Our expertise and knowledge leadership in information technology can, in fact, be a huge contributor in initiating and nurturing this kind of forward integration. With internet and mobile connectivity now available across the spectrum, what stops us from using the booming e-commerce platform to solve the distribution and infrastructure handicap that our farmers and small entrepreneurs face? Nothing, in fact, no agro-entrepreneur should underestimate what the internet and e-commerce can do for them. Additionally, one can never overemphasize the importance of growing the ecosystem by nurturing small entrepreneurs. This will not only give us inclusive growth, but is also a step towards assured business success and our own personal experience with coffee is a clear example. We tend to associate entrepreneurial success and employment generation with large IT companies. But the agri-sector can do it too. Hundreds of companies can come out of India, not just creating wealth, but creating employment at the bottom of the pyramid. India's growth story has just started and this sector can not only meet the domestic demands, but also fulfill the employment needs of an emerging country. Using technology, both home-grown as well as imported, can give the farmers and the small entrepreneurs a substantial boost to progress and compete aggressively on the world stage.

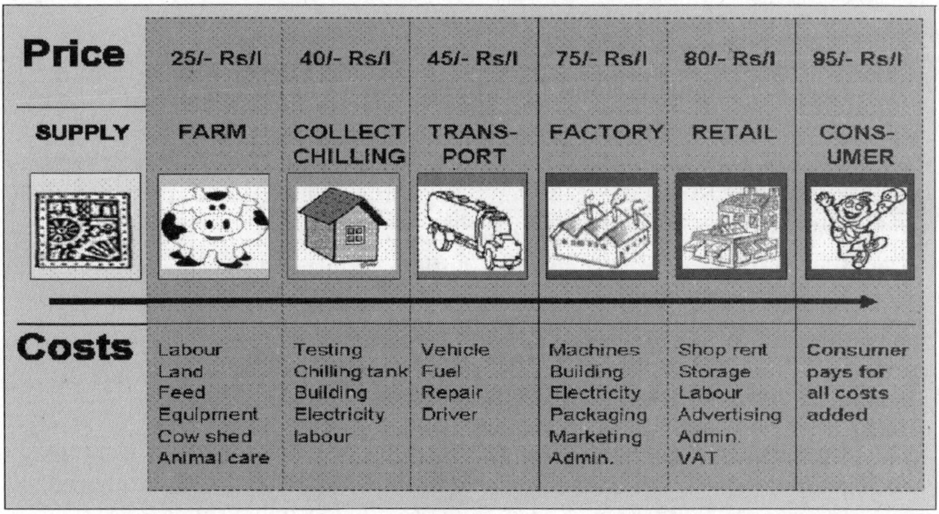


**Figure 6.1:** Value chain for dairy products: Product 1liter ultra heated fresh milk package

## Gaps

1. **Inefficient price signals**: The government has been buying almost one-third of all rice and wheat produced in India through the FCI, but in other kinds of grains, fruits and vegetables (both being highly perishable), the role of the government is limited. This leads to MSPs being ineffective as both price signals and as insulators from the perspective of the larger agricultural population.

2. **Limited reach of mandis**: This procurement system has failed to cover the entire country evenly (back of the envelope calculation suggests that on an average, a farmer needs to travel 12 kms to reach the nearest mandi and more than 50 kms in NE India) while according to the recommendations by national farmers commission, availability of markets should be within a 5 km radius.
3. **Too many intermediaries, information asymmetry:** The above mentioned problems have led to formation of long marketing channels, with multiple intermediaries, adding to the woes of the producers of perishable agri goods. These intermediaries have led to a cost escalation of ~250% (over the cost of production) and have exacerbated the existing information asymmetries in agriculture, especially for non-MSP crops.
4. **Inadequate infrastructure for storage:** The planning commission (presently NITI Ayog) has recently estimated the gap between agri-warehousing supply and demand at 35 mn MT. Currently, public sector agencies like the FCI, Central Warehousing Corporations (CWC) and the various State Warehousing Corporations (SWC) have a storage capacity of 71 mn MT, while the private sector has close to 25 mn MT. To put the scarcity in perspective, food grain stocks held only by the government was 80 mn MT according to the FCI.
5. **Skewed distribution of capacity:** Skewed distribution of this capacity is another issue, with North India having access to 60% of the total storage infrastructure.
6. **Lack of cold storage infrastructure:** India's current cold storage capacity at 25 MT is barely sufficient for 10% of fruit and vegetables produced in the country.
7. **Lack of collateral management options:** Collateral management refers to financing of agricultural goods stored at warehouses, and is estimated to be a ~ Rs 3,500 cr opportunity by industry sources.

## Emerging Solutions through Entrepreneurship

1. **Comprehensive agriculture logistics solutions:** Entrepreneurs can provide integrated post-harvest management solutions and they have already entered the space to fill these gaps apart from providing warehousing services.
2. **Integrated cold chain solutions:** Market players can provide customized solutions for cold storage and refrigerated transportation across India for fresh and frozen commodities including specialized refrigerated storage, warehousing, transportation, distribution and logistics.
3. **Integrated cold chain solutions:** Entrepreneurs can provide customized solutions for cold storage and refrigerated transportation across India for fresh and frozen commodities their services include specialized refrigerated storage, warehousing, transportation, distribution and logistics. Integrated cold chain logistics comprising post harvest transport, cold storage,

processing, and supply through refri-trucks to the distribution center and retail store. Besides areas of contract farming and agro processing, working on improving income realizations for small farmers through yield improvements, productivity increases, and consistent produce pricing.

4. **Alternate market places:** Entrepreneurship can provide a way to bypass the long chain of intermediaries by directly connecting buyers and sellers of agricultural produce and allied services, *via* a web and mobile based information exchange platform. This is a B2B (business to business) model and aims to connect all stake-holders in the supply chain: from farmers, to buyers, to suppliers of services like labor and transportation.
5. **Reducing the information asymmetry:** Riding on the high mobile penetration in rural India, entrepreneurs are working on the problem of information asymmetry for agricultural producers, by making personalized agricultural market information available to the farmers at minimal costs, through a mobile based service. TCS' mKRISHI platform offers personalized advisory services to farmers, via mobile phones (SMS and IVR), enabling them to access important information on pesticides, fertilizers, soil and water conservation, and improving access to markets for them.
6. **Innovative ICT tools for supply chain management:** Logistimo is a hosted web service for supply chain management, which can be accessed *via* basic mobile phone and web browsers, which makes it uniquely suitable for rural markets. It is a configurable service which offers customers the ability to capture and share data in a simple, low-cost way, empowering them to make better logistics decisions.
7. Although it's still early days, these solutions should lead to better supply chain management in Indian agriculture, reducing inefficiencies and increasing farmer realizations, as well as curbing food waste.

## ICT Applications through Entrepreneurship for Agri-Business Supply Chains

The global food industry has undergone significant structural changes in recent years that have created opportunities for smallholder farmers in developing nations. The inclusion of these smallholders in agri-business supply chains offers significant opportunities as well as challenges. ICTs can aid smallholders in taking advantage of opportunities and mitigating some of the challenges.

The global food industry, with over US$ 4 trillion in annual retail sales, comprises agri-businesses of varying sizes. The largest are multinational corporations that operate internationally. In this module, agri-business refers to a wide range of opportunities for entrepreneurs.

- Retailers such as supermarkets or convenience stores.
- Food processors and manufacturers.

- Input suppliers who produce fertilizer, seed, pesticide, irrigation equipment, and farm machinery.
- Producers that may be smallholder farmers, cooperatives, or corporate farming entities.
- Wholesalers, traders, and other intermediaries in a supply chain that connect retailers or processors to producers.

Changes in the informal supply environment have accompanied changes in the broader industry. The entry of large, private, often international agri-businesses from the formal economy has caused fragmented, informal suppliers to consolidate and formalize. To meet supply requirements arising from changing demands, agribusinesses often prefer to source through lead farmers or farmer cooperatives or directly from individual farmers. The alternative — purchasing from wholesale markets — can pose difficulties, be inefficient and most important, cost more. When sourcing from wholesale markets, agri-businesses have little quality control, face uncertainties in supply and price, and lose the ability to trace products (which consumers increasingly demand). Direct procurement and improvements in production, transport, and supply-chain technologies make it possible to source competitively from vast numbers of suppliers and increase the relative importance of factor costs such as labor and raw materials. Companies looking to economize move production to places where factor costs are lower, which presents an enormous opportunity for farmers in developing countries.

Commercial agri supply chain refers to a supply chain in which a private agri-business is sourcing agricultural produce from farmers or selling products to farmers in accordance with a profit-seeking business model. **Supply chain** typically refers to the set of buy-sell interactions as goods flow from raw materials through production to the final retailer where consumers can buy them. **Value chain** generally refers to the whole ecosystem of players involved moving from the retailer backward to the producer. Although participation in commercial supply chains presents an opportunity for smallholders to attain higher incomes and reduce poverty, these outcomes are not certain unless other important factors are addressed. For example, actual income changes depending on the crop, the time needed for farmers to learn to produce the crop more efficiently, and the quality and other standards required. Changes in income may not be sustainable unless accompanied by improved practices such as postharvest handling or risk management.

Market forces do not themselves guarantee smallholders' inclusion in modern supply chains. When possible, companies might seek to source from larger producers, who can deliver economies of scale, often are better educated, and typically also have better access to finance. Including smallholders can present significant challenges for both the agri-business and smallholder, but a strong business case can be made for both sides to work together. Smallholders can have distinct competitive advantages in certain situations. Compared to smallholders, large suppliers have greater market reach and multiple options to sell produce, so it can be riskier to source from them. It may also be less risky to source from numerous producers distributed across a wider geographic area, which can reduce systemic vulnerability to floods, droughts, and pests. Uncertainty about prices and quantities is reduced in the short term. Smallholders might simply have access to better land or other resources, and they are often more likely to follow the labor-intensive management practices required for higher-quality outputs.

With smallholders under contract farming, production can adapt more rapidly to market demand. Government or donors may offer incentives to include smallholders. On their side, smallholders can earn higher incomes. Participation also reduces their uncertainty as to who will buy at harvest and how much they will pay. Linking to a modern supply chain might be an important mechanism to address smallholders' lack of credit, inputs, extension services, and marketing resources. Smallholder inclusion is challenging for agri-business because interacting with a large group of small suppliers implies high administrative and transaction costs (for example, in controlling quality, maintaining traceability, or ensuring adherence to certification standards).

The agri-business may have to supply physical assets (land, machinery, inputs) and information (on management practices and postharvest practices) for smallholders to produce the quantities and qualities required. Farmers may have to learn how to grow new crops or obtain more costly inputs. Farmers may not honor standing agreements with agri-business at harvest, especially if the spot market offers higher prices than the company; the agri-business may not honor its commitment to purchase from farmers in case the situation is other way round. In short, a supply-chain relationship between agri-businesses and smallholders is a complicated partnership with difficult requirements on both sides. Procurers need an agreement that will provide them with the right product mix, items that meet safety standards and are traceable, items of the right quality, timely delivery, and a cost-effective arrangement. Farmers require market information, extension services, risk-management capacity, financial services, and supporting physical infrastructure services such as roads, storage, power, and telecommunications.

Modern ICTs and their applications significantly affect smallholders' inclusion in commercial supply chains. ICT applications (hardware and software), guided by business logic, can foster smallholders' inclusion by making the following interventions in the supply chain: reducing costs of coordination (collection of production, distribution of inputs, and so on); increasing transparency in decision making between partners; reducing transaction costs; disseminating market demand and price information; disseminating weather, pest, and risk-management information; disseminating best practices to meet quality and certification standards; collecting management data from the field; and ensuring traceability. Such interventions have been driven by the private and public sector. Their slightly different focus and resource base influence the kinds and the sustainability of ICT applications they propound. The private sector views its supply-chain relationships as a competitive advantage. The ICT applications develop to engage with the supply chain and provide information services are typically exclusive to its suppliers. Larger agri-businesses are also likely to have the scale and resources to deploy more expensive, commercially available ICT solutions within their supply chain. These interventions, if supported by a viable business model, are likely to be sustainable, but agri-business-driven interventions may not necessarily focus on smallholders. The public sector, donors, and civil society typically see the inclusion of smallholders as a public good. Their intervention in the supply chain is therefore focused on inclusion. The applications they create and develop are less likely to be exclusive. Because they have usually been designed to be specific to particular projects and used only once, public-sector interventions are unlikely to be easily generalizable to other contexts. Sustainability of donor-supported efforts has therefore been more uncertain, though future designs are expected to incorporate learning from previous experiences to enhance sustainability.

## Conclusion

Agri-business refers to the various businesses involved in food production, including farming and contract farming, supply of seeds, fertilizers, agro-chemicals, farm machinery, wholesale and distribution, processing, marketing, and retail sales. Agriculture has evolved into agri-business and has become a vast and complex system that reaches far beyond the farm to include all those who are involved in bringing food and fiber to consumers. It represents three part system made up of:

1. Agricultural input sector
2. Production sector and
3. Processing-manufacturing sector

The success of agri-business is dependent on proper planning, research, location, layout and size, climatic condition, etc. The rising food prices, increasing unemployment in rural areas, wastage of food grains due to lack of storage and demand for nutritious food etc. creates a tremendous opportunity for agricultural entrepreneurs particularly in developing nations across the world. Agricultural entrepreneurs will strive to develop agrarian prosperity while also focusing on sustainable development. They will be useful in transforming rural communities in to a contributory resource for economic development. Market liberalization and increasing consumer demand in both developing and developed countries offer attractive opportunities for agricultural producers/traders. Trade in fishery, dairy, fruit and vegetables and organic products extend altogether new options for businesses. This presents major challenges in the form of state-of-the-art technologies and infrastructure for supply chain. Through supply chains, producers in developing and emerging economies can access market information and knowledge to hone their value-added activities to both domestic and international customers.

The issue of food wastage is central to India's efforts in combating hunger and improving food security. While focus has been on improving production, reducing food supply chain losses remains a relatively unaddressed problem till very recently. It is hard to figure how much food is lost and wasted in India everydaydue to lack of adequate infrastructure, however, a 2011 report by a UN body, FAO, puts wastage in fruits and vegetables as high as 45% of produce (post-harvest to distribution) for developing Asian countries like India. In India, a large part of the agri supply chain ecosystem is either in the public sector, or strongly linked to it. The Indian government attempts to insulate the cultivator from price fluctuations by procuring their produce at minimum support prices (MSPs), decided by the commission for agricultural costs and prices after analyzing the costs of growing a particular crop. The 7500+ Agricultural Produce Marketing Committee (APMC) mandis provide a marketplace for the transaction and the Food Corporation of India (FCI) plays the role of the buyer, storing the procured produce in the relevant warehousing corporation's warehouses. Ultimately, this gets distributed through the public distribution system (PDS) shops and reaches the consumer. For non-MSP crops, the producer is dependent on the traditional private channels to market her produce.

Agriculture is a state subject and a large part of investment as well as regulatory progress is happening at the state level. Till very recently, regulatory barriers had constrained the development of storage and processing infrastructure but measures like inclusion of agri-warehousing under priority sector lending by RBI, subsidy schemes, tax incentives and the Warehousing Act (which will promote negotiability of warehousing receipts) have helped private players take an active interest in the same. The private entrepreneur guarantee scheme is one such initiative to incentivize private investment for construction of warehouses by private entrepreneurs, with an FCI guarantee to hire them for 10 years, assuring a fair return on investment by the entrepreneur.

■■■■

*Chapter 7*

# ENTREPRENEURSHIP OPPORTUNITIES IN AGRI RETAILING

**Baljeet Singh**

*Doctorate Scholar, GH Patel Post Graduate Institute of Management, Sardar Patel University, V V Nagar, Gujarat*

Retailing is an essential function of agricultural sales and distribution as it is the nearest function of end consumption. Retailing is necessary for a number of reasons including difference in buying lot sizes as most families are nuclear these days, product perishability, carrying cost of the inventory, most of the times retailed product is visible in a better way compared to wholesale product. A small family buying in bulk even if the product is not perishable, blocks the money for certain period of time which could have been put to other uses otherwise. In a nutshell, these days most buying habits are addressed by the retailers. Retail business in India roughly accounts for 22% of its GDP.

In January 2012, India approved reforms for single-brand stores welcoming anyone in the world to innovate in Indian retail market with 100% ownership, but imposed the requirement that the single brand retailer sources 30% of its goods from India. On 7 December 2012, the Government of India allowed 51% FDI in multi-brand retail in India subject to approval by the individual state where the FDIwants to set up operations. Further, a restriction on location of outlets exists, with only cities with a population of one million or more allowed to have FDI-infused multi-brand retail formats. In past three years there has been a series of allowing and disallowing of FDI in multi brand retail. In 2016, 100% FDI in processed food (multi brand) is allowed.

On the agri input side there are around 2.82 lakh retailers in India, almost one for every two villages. On the output side, number of super markets are around 5000, total number of individual retailers-11.2 million (2/3 of total), exclusive food retailers-3.7 million. However, organized food retail is only 4-5% of the total; rest 95-96% is unorganized. History of retailing shows that both unorganized and organized retail can co-exist well without significantly eating into each other's market share. In fact, it will lead to competition and fairer markets. Per capita retail space is lowest in India. Entrepreneurship opportunities in agri retailing will be discussed under three heads- agri input retailing, agri output retailing and retailing of processed food.

## AGRI INPUT RETAILING

Agri input retailing is a thriving business particularly in rural India. It is estimated that a village having one thousand households involved in agriculture buys agricultural inputs to the tune of Rs One crore in a year. Many agri input retailing formats have been prevalent in India. Some of these are discussed below.

### Chaupal Sagar

ITC launched India's first rural retail mall by the name of Chaupal Sagar. It was designed as one stop solution for the farmers and other rural people. Important feature is that ITC also does the procurement of farm produce from farmers at such centers and pays them instant cash. The farmers can use this cash for their purchases. Alternatively the farmer can order the products sitting in his village through the sanchalak who will then transmit the combined order for the village (include satellite villages) to the ITC representative, who will make it convenient for the goods to be delivered in the village depending on the bulk and logistic arrangements. Usually the logistic is available because ITC uses the same vehicles for procurement of farm produce from village farmers.

### Godrej Aadhar

It has been started as a part of Godrej Agro Vet Limited, now taken over by Future Group, Godrej Aadhar stores were projected as *Aadhaar - Khushiyon Ka, Khushhali Ka* meaning the base for happiness and prosperity.

### Tata Kisan Sansar

This started with Tata Kisan Kendra as an agri input store but later services and other products were also included to call it **Tata Kisan Sansar**. It is an endeavor of Tata Chemicals Limited. The major objective of this format is to help farmers and rural consumers to create value and improve relationship.

### Hariyali Kisan Bazar

It is the initiative of DCM Shri Ram Consolidated Limited. These have been projected as one stop shop for farmers as well as rural consumers selling from agri inputs including services to FMCG. The Hariyali Kisaan Bazaar chain, seeks to address key constraints of Indian farmer. It has provision of all encompassing solutions to the farmers under one roof. Each Hariyali Kisaan Bazaar center covers villages in a radius of about 25-30 kms. A typical center caters to agricultural land of about 50,000-70,000 acres.

### Mahindra Krishi Vihar (MKV)

It was a brand by Mahindra Shubhlabh Services Private Limited, a subsidiary company of Mahindra and Mahindra Limited. An MKV aggregated the factors of agricultural production through solutions specific to crop, region and market. It established the market linkages for balanced commodity supply chain and also offered a wide range of products and services to improve agricultural productivity. The services included bank loans with minimum paper work and training farmers on proper products and equipment use. Some of the bigger equipments were hired out also.

### Champion Agro World (CAW)

CAW is a venture of Champion Agro Limited, Rajkot, Gujarat. It is an innovative chain of rural agricultural malls set up first in Gujarat in 2005. It is a serious effort aimed at empowering farmers and meeting the needs of rural households by providing access to agricultural products, services and consultancy. Established in the countryside, the agro mall offers:

1. Quality inputs (fertilizers, seeds, pesticides, farm equipments, veterinary products, animal feed, irrigation equipments etc.).
2. Agronomic services with teams of farm staff workers and agronomists like soil testing, crop inspection, weather forecasts etc. They do this on Champion Agro bikes.
3. Financial services like farm credit, life insurance, medical insurance, crop loan and term loan.
4. Access to output markets by helping farmers with buyback opportunities, commodity trading etc.
5. Information (new development and research in agriculture, government schemes and subsidies, market prices etc. It brings out its monthly magazine in local language).

Each CAW centre caters to communities within a 25-30 km periphery and impacts the life of about 20,000 households. The Champion Agro centers truly work in the area of rural development through its retail agro centers strategy. CAW is a pioneering micro level effort, which is creating a far-reaching positive impact in bringing a qualitative change and revolutionizing the farming sector in India. The company plans to scale up the existing 34 agro centers in Gujarat to 400 centers in India.

### Opportunity in Agri Input Retailing

Opportunity in agri input retailing is huge. India is the second largest consumer of fertilizer in the world next to China. There is a large market for suppliers to sell the finished fertilizers as well as raw materials. It is also a financially credible market. The size of the Indian Agrochemical industry is estimated to be around USD 4.3 billion. Only 25% of cropped area in India receives pesticides. There are 6.38 lakh villages in India but the total number of agri input dealers is only 2.82 lakhs which brings us to the conclusion that on an average for every two villages there is one agri input dealer. In other words, in general half of the villages do not have a dealer. On the other hand Govt. of India is mulling on the idea of agriculture graduation compulsory for starting agri inputs business (already being implemented in some states). This however is good for agri graduates as it will remove a fair amount of competition. With the increasing health consciousness among people with knowledge economy, the demand for organic foods is highand so is the demand for organic agricultural inputs. Around six lakh tractors are sold every year in India. Commercial seed market is worth Rs.4050 crores (including hybrid vegetable seeds of Rs.180 crore) which constitutes 4% of world market.

## Starting your Input Retail Business

To begin with, you should have thorough knowledge of plant protection and crop production principles. Select a site which has approach roads converging from a large number of villages. A crossing or a junction of roads would be suitable. Outskirts location will save farmers' travel time inside city traffic as well as will be cost-effective rent wise. Supplies can also reach easily as trucks are not allowed inside many cities during day time. MANAGE imparts a two months course on Agri Clinics and Agri Business Centres (AC&ABC) for unemployed agriculture and life science graduates, holistically covering all aspects from starting of an AC&ABC to operationalizing and managing, even assisting for bank loan financing. A necessary input has to be a service to the farmers (paid) which can range from soil testing services to agronomic advisory.

A shop space of around 300 sq. ft. is sufficient to start the agri input retail outlet. The front part can be used for display while the back part can be converted to office for meeting with company officials or progressive farmers. Confidential discussions can also be held in this part. A godown space of around 400 sft is also required, particularly must if you are dealing with fertilizers.

## Initial steps

Choose a suitable name for your store and get it registered. Also get shop establishment registration certificate from local body. Next step is to get the licenses of sale and storage of seeds, fertilizers and agro chemicals. Principal certificates (for distribution) will be given by the companies concerned. Approach a few good company officials, secure their distributorship looking at the products, brands and local demand in the area by paying the security deposit and other necessary documents. Though it will be difficult in case of MNCs and big companies as they ask for past few years turn over, balance sheet etc. It will be good to hold some farmer meetings in the nearby villages give free samples, purchase vouchers etc to create awareness about the store in surrounding villages. At the same time it is also necessary to have contact with around 30-45 small retailers at taluka, village level for larger reach and penetration to supply them as distributor. As a first step, you can have a distributorship of five companies with a bifurcation of one vegetable seeds company, one field crops seeds company, one agrochemicals company (pesticides, fungicides, weedicides) one micronutrients and nutrient supplements company and one liquid biofertilizers company. An example is Bejo Sheetal (for vegetable seeds), Mahyco (for field crops seeds), Rallis (for agro chemicals), Aries Agro (for micro nutrients) and Agriland Biotech (for liquid biofertilizers). In fact many companies sell a wide range of agri inputs and some even sell all categories of agri inputs. Though companies will also look at the fact that if they already have a distributor in your area. You should also be a retailer to some other distributors, around 15 in number to have a diversified portfolio of products. It should be borne in mind that retailers get higher per unit margin as compared to distributors.

## Store arrangement

While arranging your store, there are few things to be kept in mind. While bigger micronutrient, fertilizer bags can be arranged outside the store, the items which you want to dispose off early (near expiry) and those with higher margins should be placed at eye level in the store shelves. The items which are demanded by farmers by name can be placed higher or lower than the eye level. Within a shelf the smaller packings should be placed in front while the bigger ones behind. Ensure that the label with trade name is distinctly visible. Finalize an easy to find arrangement for placing the items in shelves like fungicides one side, insecticides on the other side, acaricides separately and supplements in one corner or alphabetical or season-wise or crop-wise (Table 7.1).

## Indicative product range

**Table 7.1:** Indicative product range of agro-chemicals

| Common Name | Biostadt Name | Rallis Name | Dow Name |
|---|---|---|---|
| Kasugamycin 3% SL | Biomycin | -- | |
| Tricyclazole 75 WP | Blastin | -- | Beam |
| Thiophenate Methyl 70% w/w | Roko | -- | |
| Metalaxyl 8%+ Mancozeb 64% | Sanchar | Master | |
| Macozeb 75% WP | Sparsh | -- | DM 45 |
| Hexaconazole 5 EC | Trigger | Contaf | |
| Sulphur 80% WG | Wokovit | -- | |
| Propiconazole 25% EC | Redeem+ | -- | |
| Carbendazim + Mancozeb | Bendaco | | |
| Cypermethrin | Auzar 25 EC | Ralothrin | |
| Cartap Hydrochloride | Dartriz | | |
| Thiamethoxam 25% WP | Evident | | |
| Triazophos | Fullstop 40EC | | |
| Fipronil 0.3% GR | Janbaaz | | |
| Lambda Cyahalothrin | Kozuka 2.5 EC | Topgun | |
| Quinolphos 25 EC | Krush | | |
| Hexithiazox 5.45% EC | Maiden | | |

| Common Name | Biostadt Name | Rallis Name | Dow Name |
|---|---|---|---|
| Profenofos 50% EC | Prahar | | |
| Alfamethrin | Stop 10 EC | Tata Alfa | |
| Chlorpyriphos | Strike 20 EC | Tafaban | Dursban 20 EC<br>Predator 50 EC |
| Acephate | Tremeor 75 SP | Asataf 75% SP | |
| Chlorpyriphos 50% + Cypermethrin 50% | Ulka 505 | Cyclone 505 | Nurelle D 505 |
| Imidcloprid 17.8% SL | Ultimo 200 SL | Tata Imida | Imiden |
| Acetamiprid | Wapkil 20 SP | | Pride |
| Emamectin Benzoate 5% SG | Bioclaim | | |
| Imidacloprid 70% WG | Edan | | |
| Glyphosate 41% SL | Brake | | |
| Clodinafop Propargyl 15% WP | Machis | | |
| Pretilachlor 50 EC | Remove | | |
| Sulfosulfuron | Saksham | | |
| Paraquat Dichloride 24% SL | Parq | | |
| Atrazine 50% WP | Dhwansh | | |
| Metsulfuron Methyl 20% WG | Metgrip | | |
| Dimethoate | -- | Tafgor | |
| Ethion | -- | Tafethion | |
| Ethion+Cypermethrin | -- | Nagata | |
| Fenpyroximate | -- | Phenoxypyrazoe | |
| Fenvalerate | -- | Tatafen | |
| Malathion | -- | Malataf | |
| Metalaxyl | -- | Metataf | |
| Bromodiolon | -- | Ratox | |
| Carbendezim | -- | -- | Bengard 50 WP |
| Myclobutanil | -- | -- | Systhane |
| Cyhalofop butyl | -- | -- | Clincher |
| Oxyflurofen | -- | -- | Goal |
| Weedall | -- | -- | Glyphosate |
| Spinosad | -- | -- | Success<br>Tracer |

**Note:** Company and brand names are only for the knowledge of the reader and should not be taken as endorsement by the author.

**Table 7.2:** Indicative product range of plant nutrient supplements, plant growth regulators, micronutrient mixtures, multiplexes etc

| S. No. | Product | Active Ingredient | Company |
|---|---|---|---|
| 1 | Humicil | Humic Acid | Dow |
| 2 | Miraculan | Tricontanol | Dow |
| 3 | Swarnim | High Grade Organic manure | Abellon |
| 4 | Ecozyme | See Weed Extract | Ajinkya |
| 5 | Powerkin | Cytokinin | Crop Life Sciences |
| 6 | Demsin | Chloromequate Chloride | Crop Life Sciences |
| 7 | Multiplex Annapurna | Organic matter fortified with neem, castor, pongamia cake & vermicompost | Multiplex Group |
| 8 | Emerald GG 5 | Micronutrient mix (Fe,Mn,Zn,Cu,Mo,B) | Pramukh Agriclinic |
| 9 | Nectar | Liquid Micronutrient Mix | Pramukh Agriclinic |
| 10 | Vardent | Foliar Liquid Micronutrient Mix | Pramukh Agriclinic |

**Note:** Company and brand names are only for the knowledge of the reader and should not be taken as endorsement by the author.

**Table 7.3:** Indicative product range of chemical fertilizers

| S.N. | Product | Active Ingredient | Company |
|---|---|---|---|
| 1 | Sardar Urea | Urea | GSFC |
| 2 | Sardar Ammonium Sulphate | Ammonium Sulphate | GSFC |
| 3 | Sardar DAP | Di Ammonium Phosphate | GSFC |
| 4 | 12:32:16 | NPK | GSFC |
| 5 | Uttam Veer | Urea | Chambal Fertilizers |
| 6 | Uttam DAP | Di Ammonium Phosphate | Chambal Fertilizers |
| 7 | Uttam MOP | Muriate of Potash | Chambal Fertilizers |
| 8 | Uttam SSP | Single Super Phosphate | Chambal Fertilizers |
| 9 | Mahadhan Bensulf | Elemental sulphur 90%+ Bentonite Clay 10% | Deepak Fertilizers |
| 10 | Mahadhan Amruta | Water Soluble Fertilizer | Deepak Fertilizers |
| 11 | Mahadhan Chetak | DAP | Deepak Fertilizers |
| 12 | Mahadhan Potash | MOP | Deepak Fertilizers |
| 13 | Mahadhan Power | Nitrophosphate | Deepak Fertilizers |
| 14 | Bhoodhan | Granulated NPK | Deepak Fertilizers |
| 15 | Mahadhan Sulphate | Ammonium Sulphate 20.6% + Sulphur 24% | Deepak Fertilizers |
| 16 | Mahadhan Super | Single Super Phosphate | Deepak Fertilizers |

**Note:** Company and brand names are only for the knowledge of the reader and should not be taken as endorsement by the author.

**Table 7.4:** Indicative product range of organic pesticides, bio-pesticides, bio-fungicides, bio-control agents, biofertilizers etc

| S. No. | Name of the product | Active Ingredient | Manufacturer |
|---|---|---|---|
| 1 | Shakti | Growth Promoter | SRISTI Labs |
| 2 | Prayas | Flower Drop Preventer | SRISTI Labs |
| 3 | Rakshak | Anti sucking pests (Cotton) | SRISTI Labs |
| 4 | Krushak | Anti Leaf Curl | SRISTI Labs |
| 5 | Suraksha | Anti Termite | SRISTI Labs |
| 6 | Shastra | Anti Aphids, White Fly, Heliothis | SRISTI Labs |
| 7 | Prahar | Anti Mealy Bug | SRISTI Labs |
| 8 | Sarvatra | Anti Sucking Pests | SRISTI Labs |
| 9 | Monitor | *Trichoderma Viridae* | Agriland Biotech |
| 10 | Yorker | *Pacielomyces lilacinus* | Agriland Biotech |
| 11 | Biosoft | *Bauveria bassiana* | Agriland Biotech |
| 12 | Vertisoft | *Verticillium lecanii* | Agriland Biotech |
| 13 | Metasoft | *Metarhizium anisoplae* | Agriland Biotech |
| 14 | Sudozone | *Psuedomonas flourescens* | Agriland Biotech |
| 15 | Vanguard | *Azadirachtin* | Agriland Biotech |
| 16 | Rhizocell | *Rhizobium* | Agriland Biotech |
| 17 | Azospee | *Azospirillum* | Agriland Biotech |
| 18 | Biomobile K | Potash Mobilizing Bacteria | Agriland Biotech |
| 19 | Biosmart Zn | Zinc Solubilizing Bacteria | Agriland Biotech |
| 20 | Acetocell | *Acetobacter* | Agriland Biotech |
| 21 | Biofield Liquid (NPK Power) | *Azotobacter chroococcum, Bacillus polymyxa, Ascophyllum nodosum & Sargasum wighitti* (sea weed extracts), Humic acid | Agriland Biotech |

**Note:** Company and brand names are only for the knowledge of the reader and should not be taken as endorsement by the author.

## Services: soil analysis and amendment recommendation

Portable and handy instant soil testing kits can be procured from Transchem Agritech, Vadodara. One kit can carry out 100 tests at very economical rates. For second kit only refilling of reagents is required which is still cheaper. Farmers will also be taught about soil sampling technique because right sampling method is the foundation for right soil test results. The kit can measure 10 parameters namely pH, organic carbon, nitrate nitrogen, ammoniacal nitrogen, phosphorous, potassium, calcium, magnesium, sulphur and electrical conductivity. In the peak testing season (non crop season, two months after fertilizer application), field boys will visit the farmers on bike along with the kit and practically demonstrate to the farmers proper soil sampling methods from the field. Farmer meetings will be done on the crop fields of those farmers who got the soil tested and followed proper recommendation. The crop vigour, health and probable yield will be compared with adjacent fields where soil testing was not done and amendments not followed. Later harvest days can also be taken on such fields.

In order to attract more farmers, instant pH testing in soil sample at the agri input store by the farmer, by pH meter will be done free of cost. It will indicate acidity or alkalinity of the soil and proper recommendation (agricultural lime for acid soils and gypsum for alkaline soils) will be give to the farmer. The results are visible within one crop season.

*Investment*

**Table 7.5:** Fixed costs in setting up a small size agri input retail shop

| S. No. | Head | Cost (Rs.) |
|---|---|---|
| 1 | Land and building 11 months advance rent for 20 ft X 10ft shop @ Rs 3000 per month | 33000 |
| 2 | Security deposit of 10 companies @ Rs. 25,000 each | 250,000 |
| 3 | Electricity connection | 6500 |
| 4 | Furnishing | 150,000 |
| 5 | Computer, printer, fax & internet connection | 50,000 |
| 6 | Inaugural function (with lunch, 200 guests) | 250,000 |
| 7 | Decoration, boards, labels etc | 25,000 |
| 8 | Permits, licenses, certifications etc | 15,000 |
| 9 | Misc expenses | 10,000 |
| 10 | Bolero pick up | 475,000 |
| | **Total** | **1264500** |

**Table 7.6:** Variable costs for a small size agri input retail shop

| S.N. | Head | Year-1 | Year-2 | Year-3 | Year-4 | Year-5 |
|---|---|---|---|---|---|---|
| 1 | Agrochemicals | 1200000 | 1344000 | 1505280 | 1685914 | 1888223 |
| 2 | Fertilizers | 500000 | 560000 | 627200 | 702464 | 786760 |
| 3 | Biofertilizers | 100000 | 112000 | 125440 | 140493 | 157352 |
| 4 | Nutrient supplements, organic inputs etc | 200000 | 224000 | 250880 | 280986 | 314704 |
| 5 | Soil testing kits | 120000 | 134400 | 150528 | 168591 | 188822 |
| 6 | Salaries and wages | 72000 | 80640 | 90317 | 101155 | 113293 |
| 7 | Refreshments | 30000 | 33600 | 37632 | 42148 | 47206 |
| 8 | Electricity charges | 8400 | 9408 | 10537 | 11801 | 13218 |
| 9 | Internet charges | 3000 | 3360 | 3763 | 4215 | 4721 |

| S.N. | Head | Year-1 | Year-2 | Year-3 | Year-4 | Year-5 |
|---|---|---|---|---|---|---|
| 10 | Telephone charges | 7200 | 8064 | 9032 | 10115 | 11329 |
| 11 | Stationery | 4320 | 4838 | 5419 | 6069 | 6798 |
| 12 | Farmer meetings | 80000 | 89600 | 100352 | 112394 | 125882 |
| 13 | Field visits | 24000 | 26880 | 30106 | 33718 | 37764 |
| 14 | Fuel | 60000 | 67200 | 75264 | 84296 | 94411 |
| 15 | Marketing | 75000 | 84000 | 94080 | 105370 | 118014 |
| 16 | Misc | 50000 | 56000 | 62720 | 70246 | 78676 |
| | **Total** | **1333920** | **1493990** | **1673269** | **1874062** | **2098949** |

**Table 7.7:** Returns from a small size agri input shop

| Particulars | Year-1 | Year-2 | Year-3 | Year-4 | Year-5 |
|---|---|---|---|---|---|
| Revenue | 1627382 | 1822668 | 2041388 | 2286355 | 2560718 |
| Costs | 2598420 | 1493990 | 1673269 | 1874062 | 2098949 |
| Profits | -971038 | 328678 | 368119 | 412293 | 461769 |

Business would break even in the second year

## AGRI OUTPUT RETAIL

Agri output retail includes grains, fruits, vegetables, spices, flowers, milk, poultry, meat etc. Traditionally farmers were allowed to sell notified commodities only in the APMC market yards, sub market yards but with the advent of model APMC act, 2003 and its adoption (with convenient changes) by more than 50% of the states, the farmer can sell his produce anywhere to anybody including outside the market yard. It is also not necessary for wholesale buyers to buy from designated APMC markets. This has created opportunities for direct marketing, contract farming, and futures trading of agri commodities. It has also emerged the concept of negotiable warehouse receipt.

All this has created opportunities for individual entrepreneur to buy directly from farmer and sell directly to consumer with primary processing or certain amount of value addition in terms of packaging, hygiene etc, thus eliminating the middlemen and parting the share of middlemen largely between the consumer and the farmer. As a consequence retail chains have started collection centres nearby vegetables and fruit production areas.

If your demand is assured, then you can enter into contract farming arrangements with farmers. In most of the states, Government is a party to contract farming agreements at least to the tune that agreements need to be recorded with the Government. In Gujarat, the government has a little further role, in case any dispute arises between the sponsor and farmers in contract farming. This ensures safety to both producer and contract sponsor in case any of the two parties disobey the agreement.

In case of food grains, it is advisable to have an established shop along with a storage facility. Branding can be done with private labels. In case of fruits and vegetables, door to door vending service can be explored harping on the USP of fresh, hygienic and convenience. This assumes importance with increasing middle class, disposable incomes and cities expanding. In a scenario where both husband and wife are working and do not find time even to shop fruits and vegetables though fresh greens are desired all the time, organized door to door vending in a chiller van can do wonders. It's easy to get assured customers particularly in big cities and posh areas. Roping in an IT guy can help you set up an online ordering system. Two chiller vans can make round trips in opposite directions covering 200-300 households, twice daily, once in morning and once in evening. This will expose the customer households to supplies four times a day. This will provide that the customer get his F&V supply at almost any desirable time. This will not only ensure fresh and hygienic supplies but also reduce refrigeration needs at customer homes. You also do not need to spend on making the van refrigerated. A duct mediated automatic evaporative cooling system for perishable agro produce transport in mini trucks/vans has been devised by Food Processing Technology (FPT) College of Anand Agricultural University (AAU). The system gets activated, when the mini truck/ van starts moving and remains active as long as the vehicle is in motion and works on zero energy. The technology is available for commercialization to individual entrepreneurs at minimal cost. Same vehicle can be used for procuring fruits and vegetables from farmers' fields.

Different kinds of vegetables require different primary processing/pre-cooling treatments depending upon texture and moisture content. The cleaning, pre-cooling, grading, sorting, packing (thin film cling packaging can be tried) facility can be a 20x30 ft enclosure with sufficient water supplies. After transportation from farmers' field, they immediately need to be brought to this facility. Primary processing is done and then dispatched for sale. In case of any inventory, a zero energy cool chamber which runs without external power and can keep the vegetables fresh at 12-18 degrees for 72 to 96 hours should be used. This technology is also available at nominal cost at FPT College of AAU. This will help save costs.

## Investment

**Table 7.8:** Fixed costs in setting up a door to door fresh fruit and vegetable supply system

| S. No. | Head of expense | Amount in Rs. |
|---|---|---|
| 1 | Civil works including land | 240000 |
| 2 | Mini trucks/ vans-2 | 900000 |
| 3 | Technology | 200000 |
| 4 | Electricity connection | 6500 |
| 5 | Water source | 250000 |
| 6 | Telephone, fax, internet connection | 9600 |

| | | |
|---|---|---|
| 7 | Computer, IT related, website, software etc | 60000 |
| 8 | Furnishing | 175000 |
| 9 | Inaugural function | 200000 |
| 10 | Misc | 100000 |
| | **Total** | **2141100** |

**Table 7.9:** Variable costs in setting up a door to door supply of fresh fruits and vegetables system

| S. No. | Head of Expense | Year-1 | Year-2 | Year-3 | Year-4 | Year-5 |
|---|---|---|---|---|---|---|
| 1 | Manpower | 144000 | 161280 | 180634 | 202310 | 226587 |
| 2 | Electricity | 10800 | 12096 | 13548 | 15173 | 16994 |
| 3 | Water | 18000 | 20160 | 22579 | 25289 | 28323 |
| 4 | Communication | 9600 | 10752 | 12042 | 13487 | 15106 |
| 5 | Stationery | 11040 | 12365 | 13849 | 15510 | 17372 |
| 6 | Fuel | 110000 | 123200 | 137984 | 154542 | 173087 |
| 7 | Packing material | 14600 | 16352 | 18314 | 20512 | 22973 |
| 8 | F&V | 2700000 | 3024000 | 3386880 | 3793306 | 4248502 |
| 9 | Misc. | 200000 | 224000 | 250880 | 280986 | 314704 |
| | **Total** | **3218040** | **3604205** | **4036709** | **4521115** | **5063648** |

**Table 7.10:** Returns from a door to supply of fresh fruit and vegetable set up

| Particulars | Year-1 | Year-2 | Year-3 | Year-4 | Year-5 |
|---|---|---|---|---|---|
| Revenue | 4022550 | 4505256 | 5045887 | 5651393 | 6329560 |
| **Costs** | 5359140 | 3604205 | 4036709 | 4521115 | 5063648 |
| Profits | **-1336590** | **901051** | **1009178** | **1130278** | **1265912** |

The business would break even in the second year.

## RETAILING OF PROCESSED AGRI PRODUCE

### Organized and Unorganized Retail

Retailing of processed food can be classified into two categories i.e. organized and unorganized. Unorganized retail includes mom and pop stores, kirana stores, small provision stores, neighborhood stores etc. while the organized sector includes convenience stores, super markets, hyper markets etc.

### Big Opportunity in Processed Food Retailing

In recent decades, with the opening up of the economy, disposable incomes particularly that of middle class have increased and there is proportionately increased spending on processed food given the fact that there are more and more working couples, people working at odd hours. Amount of travel has also increased. People are spending more time outdoors. The affinity of the younger generation towards canned food, fast foods, and convenience foods has increased. Overall, there is also a marginal increase in the processing of fresh foods. With the increased pollution and heavy use of chemicals in agriculture, there is advent of new diseases and scales of infirmities have increased. This has given an impetus to the demand of protective foods. Increased awareness, educational standards and health consciousness has put on a heavy demand for hygienic food and out of season fruit products. All these kinds of foods require at least some amount of processing. The average Indian family spends about 35% of its income on food. Food retail industry is today valued at Rs. 15 lakh crores in India.

Looking into the big opportunity, many Indian corporate houses and MNCs have stepped into the processed food retailing sector. You may want to start anything from a kirana store to a super market, but in the beginning it is better to start with a convenience store which could be more of a neighborhood store but located in the city (not inside the residential area), although an entry point to big residential complex is the right place to start your convenience store. It is advised to stick to regional business in first few years of operation. Examples are Granary (Gujarat), Nayantara (Assam). The take off merchandise should include cereals, staples, breakfast foods, snacks, spices and condiments, milk products, FMCG products, baby foods, specialty foods, processed fruit products like juice, nectar, jam, jelly, pickle, ketchup, sauce, puree, sharbat, banana chips, preserve, marmalade and processed plantation products like tea, coffee, spices, condiments, dry fruits etc, processed vegetable products like potato chips, dehydrated onion, garlic paste, ginger paste, edible oils, drumstick powder, FMCG products can also be included.

In a sense, in order to appeal to all classes of the society, retail stores would have to identify with different lifestyles. The provision of home delivery to nearby areas subject to a minimum amount of sale can be thought of with additional delivery service charges.

## Critical Success Factors in Running a Food Retail Store

A number of important things are to be kept in mind before starting your retail store. These includeon a **store level basis** - location, ease of shopping experience, store layout, size of store, overall price position, cross category promotion, credit, atmosphere, entertainment (particularly for children), service experience, friendliness of employees, store brand equity, trust, emotional identity with store and effective management of supply chain including vendor management. On a **product level basis,** these are category such as performance, quality, price, selection, product warranty, bundle pricing, maintenance agreements, return policies, discounts, product display, product brand equity emotional connection with store brands and trendiness and effectively managing the product losses. Generally, shoppers have just one or at the most two compelling reasons for patronizing a store. This reason would be the overriding criterion that would cause a shopper to go in a store.

## Categories

Category in retail management includes the products which can be reasonable substitute to each other. For example, biscuits in one category, now whether it is Parle G or Sunfeast or Good day, all will form a category as they act as close substitutes of each other. It is not feasible to have a complete category but it is advised to have a number of brands in each category as per consumer taste. It is also suggested to have all the impulse buying categories displayed at the eye level and, cereals, staples and other essentialities to be placed few levels above floor while children's' category items could occupy the lowest levels. In the bigger stores, you find category managers while in the smaller stores all the categories are managed by a single or a few persons. It can be easily deduced that while brand management focuses on optimization of single brand, category management focuses on the optimization of a group of brands.

It is better to first define the categories before laying out your stor then define the sub categories. The products of related categories could be placed in adjacent racks or one over the other racks. For example, bread and jam could be placed up and down; nearby each other but keep in mind that bread needs to be replaced almost every alternate day and while jam has a higher shelf life. While defining categories, products can also be classified in certain convenient ways, for example below is the classification of products for ease of convenience.

1. **Specialty**: Fruit juices, beer, soft drinks, pro-biotics, cakes etc. high assortments, good product knowledge, specifications and standards required.
2. **Traffic builder**: Milk, bread, cereals and staples (to some extent) promotion and competitive pricing required.
3. **Service:** Tea, coffee and other beverages- limited assortment and competitive pricing required.
4. **High margins:** Breakfast snacks, meat noodles etc. the communication and excellent presentation is required.

Categorization also offers many other benefits like reduction in stock keeping units (SKUs). Indirectly, it also reduces costs and improves revenues. It also reduces the time and effort in handling the products and the store. As Zeithaml (1988) found that consumers consider both monetary and non monetary costs- such as time and effort to evaluate the value of shopping at a particular store, anything that can be built to reduce time, effort and search costs can increase perceptions of value. Search costs of a shopper can be reduced by properly laying out the laying out and displaying merchandise. Use of appropriate signage can also reduce search costs.

### Store Hygiene

Hygiene holds utmost importance while dealing with food products. Store needs disinfection at least every year. This should be done after evacuating the store. Any rodent holes etc should be plugged. Cleaning of the store including behind the shelves should be done daily. There should be a minimum 1-2 inches distance between the floor and the shelf (lower stack) for easy cleaning. Putting in CCTV cameras will not only keep a watch on shoplifting but can also help detect movement of rodents. This will also keep the employees under scan.

### Sales Forecast

Many Indian food retailers do not carry out regular sales forecasts unless demanded by the supplier or before peak demand season like festivals etc. Though sales is influenced by a large number of factors but doing a sales forecasts exercise based on past data, simple surveys, taking into consideration known factors influencing sales can go a long way in better inventory management and cost reduction in the processed food retail business.

### Supply Chain Management

Supply chain starts from the **sourcing** point. As the store will have different brands in each category, you will have to deal with a number of vendors. It is always good to have a SAP system or any other good ERP system for your store added with bar coding system. While sourcing, it is important to keep in mind the PQRS. It is explained below:

P = Procuring from the right place

Q = Procuring the right quality and quantity

R = Procuring at the right rate (Price)

S − Procuring from thc right sourcc/supplier (Vendor)

While ordering, **inventory management** strategies and economic order quantities (EOQ) need to be considered. Food retailers, generally hold the inventory for three weeks. The inventory of certain products needs to be increased before special occasions like festivals or events in the vicinity. The back side of the store can be used as temporary storage while a bigger warehouse at a close distance nearby is suitable for long term storage. It is good to have own transport like mini truck/ Tata Ace etc for short distance transportation from warehouse to store. The formula for calculating EOQ is given below:

EOQ = Square root of 2AP/S

Where;

A= Demand per unit time

S=Carrying/holding cost of one unit stock for the given time period

P= Order cost

### Transportation

Though all the processed food products have a good shelf life compared to the fresh products, but still they possess some degree of perishability. It is important to have safe and covered transport. Open transport may lead to creeping in of unhygienic elements. For milk based products, cold chain transport is suggested. In most cases, the supplier himself arranges the transport, but it is always good to insist for a safe and protective transport.

## Bulk Buying and Repacking

Bulk buying and repackaging is in vogue in food retailing. But this should be done with the permission of the manufacturers (if manufacturer has a brand) and terms and conditions laid down by the manufacturers should be adhered to. If the item is just a commodity, then it will be a case of private label.

## Private Labels

Another trend in vogue these days is the private labels. These are often practiced by food retailers in the staples category wherein bulk sourcing is from local market and then goods are packed in retailer's own brand. This is also a source of high margins in the staples. This can be effectively done for commodities/ products which are abundantly available in the local area.

## Offering Space to Others

Space could be offered to others on rental both on carpet area basis and self space basis. This is another way of earning revenue from your food retail store. There are premium spaces such as front part of the store or the shelves at eye level. Accordingly, premium spaces attract a higher rental which funnels the synergies of both the store and the one who takes space on rental.

Organized food retailing is capital intensive business with a long gestation period (around five years). Investment is required in buying/leasing land, furniture and fixtures, IT systems, quality control systems, vendor development and infrastructure for warehousing storage and transportation. The sector is labor intensive also and results in significant employment generation.

## Suggested Readings

1. http://www.icra.in/Files/ticker/SH-2015-Q3-4-ICRA-Fertilisers.pdf accessed at 9.07 pm on 21.1.16.
2. http://www.kamtech.in/Ajay_Gupta.pdf accessed on 10:43 pm on 27.2.16.
3. Vision, Strategy and Action Plan for Food Processing Industry in India by Rabo India Finance, 2005.

■■■■

*Chapter 8*

# ENTREPRENEURSHIP OPPORTUNITIES IN FISHERY SECTOR

**[1]Shakti Ranjan Panigrahy and [2]Arpita Mohapatra**

[1]*Assistant Professor, IABMI, Anand Agricultural University, Gujarat*
[2]*Scientist, ICAR-CSWRI, Avikanagar, Rajasthan*

## INTRODUCTION

India is bestowed with a long coastal line of 8118 kms, 3827 fishing villages, and 1914 traditional fishing trading centers and restores various categories of fishery species in its both inland and marine resources. India is home to more than 10% of the global fish diversity and it ranks second in overall fish production and seventh in marine fish production (State of World Fisheries and Aquaculture, 2014). Out of total fish production of 9.58 million tonnes, marine resource contributes 3.44 million tonnes and inland resource contributes rest (Handbook of Fisheries Statistics, 2014). Besides that India has an estimated total estuarine area of 3.9 million hectares, of which 1.2 million hectares of coastal salt-affected lands have been identified suitable for brackish water shrimp farming. However, in shrimp culture, only 15% of area has been used for culture till now. Similarly, aquaculture resources in India include 2.36 million ha of ponds and tanks, 0.798 million ha of flood plain lakes/derelict waters plus in addition 1,95,210 km of rivers and canals, 2.907 million ha of reservoirs and that could be utilized for aquaculture purposes.

Ponds and tanks are the prime resources for freshwater aquaculture, however, only about 40% of available area is used for aquaculture currently (FAO Fisheries & Aquaculture- National Aquaculture Sector Overview-India) and today India ranks $27^{th}$ in value and $23^{rd}$ in volume in fish export and approximately 90% of traditional fishermen are below poverty line (Rao, 2012). Food is the biggest item of expenditure of Indian households. On an average about 51% of the total consumption expenditure of households was spent on food and out of them meat, egg, fish consumption lies near to 4% (CMIE, 2012). About 35% of Indian population is fish eater and the per capita consumption is 9.8 kg whereas the recommended intake is 13 kg (CSO-MFS, 2011). Fish contains omega-3 fatty acid that is good for the heart and fisheries are estimated to provide 16% of the world population's protein (CMIE, 2012). A portion of 150g of fish can provide about 50-60% of adult's protein requirement (State of World Fisheries and Aquaculture, 2014). Taking in a nutshell, there is an ample opportunity in India for fishery sector as it has both supply and demand advantage in all its spheres and its complete exploitation has yet to happen. At the same time poor post-harvest fish handling infrastructure in major maritime states is causing an annual loss of over Rs 15,000 crore to India's marine and inland fisheries sector (Business Standard Jan 28, 2013). That indicates investment on infrastructure must be done in coming years which generates ample opportunities for the fishery entrepreneur.

## Production and Consumption Variability in Fishery Sector

An agro-entrepreneur has enough possibilities to do his business in this sector by integrating both forward and backward linkages. The demand for this particular product is increasing day by day with increasing of consumer surplus but unlike demand its production is confined to some local areas. Heterogeneity is also marked for the preference of varieties of fisheries products. Here we find southern states prefer marine fish over inland fish and eastern Indian people prefer inland fish more than marine one. In India, we consume about 74% of marine fish and rest export to different countries and preference for marine fish is going down every year. It may be due to increase in demand for inland fish due to its lesser price than former. In aquaculture all carp varieties are consumed as fresh in India whereas the brackish water produce varieties like tiger prawn and shrimp are harvested mainly for export purpose. It is expected that at present there are more than 130 wastewater aquaculture units in India covering about 10,000 ha. About 80% of them are located in West Bengal, where sewage is extensively used as a fertilizer for fish pond. This manmade impounded environment creates a brackish water culture that helps for shrimp and tiger prawn production.

***Fish demand in India***

According to Singh et al. (2014), Tripura and Lakshadweep have been identified as maximum fish consuming states in India. These are followed by other states like Goa, Arunachal Pradesh, Kerala, Andaman Nicobar and Dadra & Nagar Haveli. In Arunachal Pradesh, the rural consumption (2.18 kg per person) is more than urban consumption (1.119 kg per person).

According to them, In Northern India, fish consumption is very low as compared to Southern and North-eastern regions. Punjab and Haryana have taken lead in inland fish production with a maximum yield of 48,000 kg/ha/yr. The trend in Punjab and Haryana has changed to a greater extent in recent years where farmers have adopted pisciculture instead of agriculture. It has no seasonal constraint and is economically more beneficial in comparison to other occupations.

The same Authors have an opined that in West Bengal, fish consumption in urban areas is higher than rural areas, whereas in Goa, it is reverse. In Rajasthan, Punjab, Haryana and Jammu & Kashmir, only inland fish production activities are progressing well and fish production is more than its consumption. In Kerala, fish consumption is higher in urban areas than rural areas. The state makes important contribution to marine fish production. During the last decade; marine fishermen have shown keen interest in adopting inland fish culture. In North-eastern regions of the country, Tripura is leading in fish consumption followed by Arunachal Pradesh. Thus, study of the Indian fisheries reveals that the states like Tripura, Lakshadweep, Goa, Kerala, Arunachal Pradesh and Andaman Nicobar Islands are the major fish consuming states/UT.

Assam, with a predominantly fish eating population, has potential for development of fisheries but at present its demand outweighs supply, leading to huge import from outside the state. The present annual fish production from all sources is about 1.6 lakh tonnes against a demand of 2.05 lakh tonnes, according to status report prepared by the state fisheries department. Due to improper technical education, demonstrations and research facilities in the sector, intending aqua culturists are unable to contribute much in fish production of the state. The state government has, however, realized the vast potential of natural resources and is currently implementing a comprehensive package with funding from the World Bank to develop sustainable fisheries in Assam.

## Innovation and Integration in fish Production

Brackish water aquaculture is a traditional practice in the form of bheries (manmade impoundments in coastal wetlands) in West Bengal and pokkali (salt resistant deep water paddy) in Kerala coast. During the preparation of brackish water aquaculture, bheries are dried up and sewage is allowed to enter bheri. After stabilization of effluents and after observing the plankton population, the bheries are stocked with advanced fingerlings @7000-10000/hectare. Normally multiple stocking and multiple harvesting is adopted in bheries and fish are reared for 3-5 months, depending on the growth of the fish to reach a marketable size of 250-400 grams (FAO Fisheries & Aquaculture - National Aquaculture Sector Overview-India). Pokkali is a unique tolerant rice variety that is cultivated in an organic way in the water logged coastal region. The rice is cultivated from June to early November when the salinity level of the water in the fields is low. From mid-November to mid-April, when the salinity is high, prawn farming takes over. The prawn seedlings which swim in the sea are captured and cultured by restoring backwater after the rice harvest. Sluice gates are used to control the water flow to the fields. The rice crop, which gets no other fertilizer or manure, draws nutrients from the prawns' excrement and other remnants. Since the tidal flows make the fields highly fertile, no manure or fertilizer need to be applied; the seedlings just grow the natural way.

## Fish Business Environment and Opportunities

Like brackish water aquaculture, cage/pen culture are promoted to enhance fishery production. Similarly, mariculture has also been developed by CMFRI to produce oysters, mussels and clams among sedentary species, as well as for shrimps and finfish. Marine fish is a capture based fish production but day by day inland fish production overtakes it, as marine resource is regulated by government law and it is enacted that no one can capture fish above the maximum permissible capture level. Besides that fish either migrate or go deeper to the sea level due to adverse climatic conditions, water pollution, eutrophication, hypoxia and many more factors. Still expansion of fleet capacity, technological innovation and increase in investment all led to explosive growth in the exploitation of marine fisheries. The Indian Ocean was the sole major marine fishing area to show sustained growth in past three decades. As an agro entrepreneur, production of fish and its effective utilization in an efficient way is quite necessary to fulfill the burgeoning demand in the context of increase in population, change in consumer taste and increase in consumer surplus. There is good news for the agro-entrepreneurs that fish remains among the most traded food commodities worldwide. In 2012, about 200 countries reported exports of fish and fishing products. In 2012, it represented about 10% of total agricultural exports and 1% of world merchandise trade in value terms (The State of World Fisheries and Aquaculture, 2014).

As fish is a perishable commodity and it is mainly consumed in fresh form as it is, entering processing sector is a most awaited task to be carried on. Unlike milk and sugar, cooperative sector is not developed in fishery as value added fishery products have not been accepted full-fledged in India as yet. But for export purpose, many processing units have been set up and they are mainly found in coastal region or near to the sea shore. Processing of fish has been done mainly in the form of canned or frozen form and only for the export purpose. Totally, there are 258 freezing units with a capacity of 2,170 tonnes, 23 canning units of 84.5 tonnes capacity, 131 ice-making units of 1820 tonnes, 24 fish meal units with a capacity of 419 tonnes and 297 cold storage units with a capacity of 2,03,448 tonnes are established in India (chp-1, The food processing sector in India: An Overview). Government has also amended some rules and regulation in the direction of fish processing. These days foreign equity is permitted in this sector. Fish processing project with a minimum of 20% value addition can set up as 100% export oriented units. All items can be exported freely except for silver pomfrets of weight less than 300 grams. Export of marine products is allowed only after registration of the units as an exporter with the Marine Products Export Development Authority (MPEDA). There may be a chance for agro-entrepreneur to enter into this field as till date per capita consumption of fish in India is about 5kg per annum as against the world average 16 kg per annum.

About 73% of fish has been marketed of total disposition of fish as fresh in India. That's why scope is more in this segment for the agro entrepreneur. In India 90% of workforce is in unorganized sector and out of them 65% is in agriculture (National Commission for Enterprises in the Unorganized Sector, 2007). In another report by NSSO (2007-2008), almost 99.99% labour input is in agriculture and forestry and 98.7% is in fishing in unorganised sector. That's why a major chunk of dependency in fishery sector is dwelling under informal sector and it needs to be organized. Similarly in an another study it was found that the average labour requirement in shrimp farming has been estimated at about 600 labour days/crop/ha as against 180 labour days/crop/ha in the case of paddy field cultivation (Rao and Ravindran, 2001). An agro-entrepreneur in fishery sector with innate interest and skill can create jobs particularly in rural sector which may be a tool to arrest migration. Advent of private sector in fishery sector created a positive indication for profit making and that attracts many more entrepreneurs to invest in fishery and allied sector ventures.

Agro-entrepreneurship in fishery sector creates avenues not only in production and processing sector but also in marketing sectors. No doubt marketing of the product generates revenues for further investment. Indian Tobacco Company (ITC) established aqua-choupal model in which prior fish supply chain is restricted within few middlemen who market the product in a better way with proper quality check of the product and astute market intelligence. The role of IT in this model cannot be ignored as in time communication between market and supplier of fish enhances value among all the stakeholders particularly producer, consumer and company as a whole. Similarly Reliance Delight Non Vegetarian retail outlet in the Chennai market proves better and smarter market intervention after thorough market research as approximately 95% of the customers were found to be non-vegetarian in nature and they prefer sea food in their diet which was a profitable business venture for the company. Spencer Company also penetrates in the fish marketing structure by doing franchise among the vendors. The company welcomes fish vendors and provides them space for sale to the customers. The basic objective to highlight all these models is to accustom the fishery agro entrepreneur with their business environment. In India, fishery sector is in the hands of unorganised sector. Various stakeholders like retailer (both male and female), wholesalers, mashakhores and itinerant traders are creating opportunities by marketing of the products. As fish is a perishable product, consumed regularly and large number of buyers and suppliers are there in the market; that's why it is a kind of perfect competition. It means to say in this particular business model profit can be increased by selling more to the customer or by making some Farmer Producer Organizations (FPOs) like cooperatives, SHGs and producer companies; as they have better retaining power and bargaining power as well. After lackluster performance of Matsyagandha and Versova Cooperative Society in fish doubts have been raised on the performance of cooperative in fish marketing. After 2008 due to devaluation of money and decreasing supply of marine fish due to its poor production throughout the world, export of the marine fish increased in a steady fashion in value terms and European market was considered as a better market destination for this product by looking into its market instability. In carp fish (Catla, Rohu and Mrigal), in Catfish (Particularly Magur), shrimp and tiger prawn are showing profitability for the entrepreneur in aquaculture. That's why many entrepreneurs are collecting eggs, fry and fingerlings by stocking the water flows from rivers, streams and lakes which are coming from Indian Himalayas and are using the same for aquaculture production.

Ornamental fish production and its marketing are also getting attention in India. Kerala has its own natural strength to march ahead of other states in this sector due to the conducive climate. Though acceptance of the ornamental fish is at its nascent stage yet its increasing demand day by day motivates entrepreneurs to invest on this sector. Ornamental fishes have gained popularity as it is easy to maintain them inside homes when compared to other pets like dogs and birds. USA is the major market for this particular variety of fish. Till date whatever we produce it is easily absorbed in the Indian market, but awareness creates avenues for the entrepreneur in coming days. About 80% of ornamental fishes are from fresh waters and the rest from brackish and marine waters. Most of the ornamental species are warm water tropical except some eurythermal carps like gold fish koi which are cold temperate in origin and now having a worldwide distribution. **Kavil**, being the first of its kind in India in the tropical pet fish sector, will also become the quality assurance agency and will allow its shareholders to market their fish under a brand name unique to Kerala breed fish.

## Endnote

In India fisheries are recognized in many pockets like Veraval for marine fisheries, Darjeeling for cold water fisheries, Lakshadeep Development Corporation Limited for tune fisheries, Sardar Sarovar for inter-state river fisheries, Kulleru Lake for lake fisheries and many regulated markets like Crawford markets in Mumbai, organized Howrah fish market are generating attractive market forum for the fishery agro-entrepreneur. In India some eminent fishery agro-entrepreneur had shown their reflection in the society like Late Nilu Ghosh near Kolkatta, believed to be the pioneer in modern fish seed production in the country, Ashis Sarkar, Dipak Roy and Kirpan Sarkar of North Bengal, Hari Churan Das of Tripura, Darsan Singh Tabiba near Ludhiana in Punjab etc. Hopefully it is the agro-entrepreneur who changes the business in their way and their need is inevitable as demand is growing for the fishery products continuously and the government is also emphasizing on food and nutritional security these days.

## Further Readings

1. CSO-MFS (2011). Manual on Fishery statistics. Ministry of Statistics and Programme Implementation, Government of India.
2. Datta K.S; Singh P.S; Chakrabarti. M; Biswas S & Bittu S (Feb. 2010). A Perspective on Fisheries Sector Interventions for Livelihood Promotion. Indian Institute of Management, Ahmedabad, India.
3. Handbook of Fishery Statistics (2014). Department of Animal Husbandry, Dairying and Fishery, Government of India.
4. Muralidhar Rao P. (2012). National Policy on fishing and Fishermen, Bharatiya Janata Party.
5. Singh R., Pandey, P.K., Salim S. & Krishnan M. (2014). Evolving fisheries business in India with GIS. India Geospatial Digest, 1-4.
6. State of World Fisheries and Aquaculture (2014). World Review of Fisheries and Aquaculture, 1, 10.

■■■■

*Chapter 9*

# ENTREPRENEURSHIP OPPORTUNITIES IN LIVESTOCK AND ALLIED SECTOR

**[1]Arpita Mohapatra and [2]Aditya Prasad Acharya**

*[1]Scientist, ICAR-CSWRI, Avikanagar, Rajasthan*
*[2]Assistant Professor, College of Veterinary Science and Animal Husbandry, OUAT*

## INTRODUCTION

Agriculture and livestock are two integral parts of human civilization without them the basic structure of a society cannot be built up. The day to day activities of rural and under privileged population are intermingled among themselves and their socioeconomic upliftment can only be possible by strengthening both agriculture and livestock sector. Again as far as land is concerned, it is deteriorating and degrading continuously in India and net cropped area is stagnant at 140 million hectare which is a major concern for future production. About 55% of agriculture is rain fed and agriculture is absorbing about 48% of workforce; although the annual growth in agriculture is quite abysmal. At this juncture livestock rearing and livestock production directly impacts the socio economic parameters of small and marginal farmers, their livelihood and their integration with the main stream of development. Increase in demand of milk and meat pushes the upward growth of this sector and impels infrastructure like storage, processing, packaging and transportation in various ways.

### Knowing the Sector

India has emerged as the largest producer of milk with 18.5% share in total milk production in the world. India accounts for about 4.95% of the global egg production and also the largest population of milch animals in the world, with 112.9 million buffaloes, 157 million goats and 74.5 million sheep. Egg production was around 78.48 billion eggs in 2014-2015, while poultry meat production was estimated at 3.04 MT. The per capita availability of milk in India has increased to 322 grams whereas per capita availability of poultry meat lies on 630g and for eggs it is 33 nos.

Export of animal products represents an important and significant contribution to the Indian agriculture sector. India's export of animal products was Rs. 33128.30 crores in 2014-15, which include the major products like buffalo meat (29,282.60

crores), sheep/goat meat (828.11 crores), poultry products (651.21 crores), dairy products (1,205.38 crores), animal casing (19.32 crores), processed meat (14.21 crores) and other meat (2.67 crores). The demand for Indian buffalo meat in international market has sparked a sudden increase in the meat exports. Buffalo meat dominated the exports with a contribution of over 86%. The product registered 27% growth in export during the financial year 2012-2013 as compared to the same period of last year. The main markets for Indian buffalo meat and other animal products are Vietnam Social Republic, Malaysia, Thailand, Saudi Arabia, Egypt, Arab Republic and UAE. In terms of export from India, poultry products, processed meat and natural honey recorded 7%, 28% and 11% growth respectively, during the financial year 2012-2013 over the last year. The major importing countries of these products were USA, Afghanistan, Saudi Arabia, Indonesia, Germany, Netherlands and Denmark etc. Whereas, dairy products sector registered a growth of 388%, skimmed milk powder (SMP) became the largest item of export from India, which accounts for nearly 77% of net milk and milk product exports during the year 2012-13, and Bangladesh, Egypt, UAE, Saudi Arabia and Algeria were the major importing countries of dairy products from the country during the period.

About 94% of production of eggs is contributed by fowls. During 2012-2013 states like Kerala, Uttarakhand, Bihar, and Madhya Pradesh registered highest growth in egg production. Out of total eggs produced, Andhra Pradesh shared 31% followed by Tamilnadu 17.1% in 2012-2013. West Bengal produces most of the duck eggs in the country. Namakkal in Tamilnadu is India's egg hub and accounts for substantial share of India's egg exports. Nearly 45% of meat production is contributed by poultry. Buffalo, goat, pig, sheep and cattle meat contribute only 19%, 16%, 8%, 7% and 5% respectively to the total meat production in the country. The largest producer of meat is Uttar Pradesh which produces 19.1% of total meat production in the country followed by Andhra Pradesh (15.2%) and West Bengal (10.9%). The average meat production from buffalo is highest in Uttar Pradesh which was 0.516 million tonne per year in 2012. The average meat production from sheep is highest in Andhra Pradesh at 0.199 million tonne. The production of goat meat is highest in West Bengal with 0.242 million tonnes per year followed by Uttar Pradesh, Andhra Pradesh and Maharashtra. The meat production from pig is highest in Uttar Pradesh with 0.172 million tonne followed by Bihar, Nagaland and West Bengal. The highest producer of poultry meat is Andhra Pradesh with 0.499 million tonne per year followed by Maharashtra, Tamilnadu, Haryana and West Bengal. The total wool production in the country is increasing slowly and amounted to 46.1 million kg in 2012-13. Nearly 71% of the produced wool is contributed by ram/weather followed by ewe (25%) and lamb (4%). Rajasthan is the highest producer of wool in the country which produces 30.4% of total wool, followed by Karnataka (17.4%) and Jammu & Kashmir (16.7%).

Looking in to the success of dairy and poultry sector across species and regions, the National Livestock Mission has been launched in 2014-15 with an approved outlay of Rs 2800 crore during twelfth plan. This mission is formulated with the objective of sustainable development of the livestock sector, focusing on improving availability of quality feed and fodder, risk coverage, effective extension, improved flow of credit, and organization of livestock farmers/ rearers (Economic Survey, 2015-16). In Budget 2016, 850 crore has been allotted in the name of four projects i.e. pashudhan sanjivani, nakul swasthya patra, e-pashudhan haat and national genomic centre for indigenous breeds which indicate the government's positive moves for betterment of livestock sector.

## LIVESTOCK FARMING IN DIFFERENT CORNERS

Majority of livestock rearing is in the hands of unprivileged persons, probably due to social factors, in the choice of livestock species. For example, the majority of tribal families in India own backyard poultry (mainly chicken and duck), pig keeping is common with tribals from Eastern and North-Eastern states but not with those from western states, where goat keeping is more common. Besides the tribals of the North-East, pig keeping in the rest of India is mainly with scheduled and backward classes, while goat keeping is common among Muslim and backward classes. There is variation between social groups amongst pastoralists mainly with respect to livestock species kept. For example, Bharwads in Rajasthan and Gujarat, Dhangars in Maharashtra keep sheep and goats, while the Rabaris/Raikas, in North-West India specialize in cattle and camel breeding. Social groups from backward classes, like Nats in Uttar Pradesh and Wagharis in Gujarat, specialize in breeding of buffaloes.

These days' livestock rearing is more inclined to Buffalo and goat among the farmers and entrepreneurs. Buffaloes have their own identity for quality and fat based milk and carabeef demand has been increasing in international market due to organic in nature to a great extent. Goat meat has been satisfying the internal market demand; although poultry sector is more organized and has captured major market share in India. Besides this, latter is handled by large and wealthy farmers. Looking into the demand of these species particularly buffaloes and goats; they are renamed as animals of the future for the country. While we are putting more importance on goat and buffaloes, sheep and poultry farming cannot be ignored along with India's most preferred traditional cattle farming. Sheep farming is much more advantageous than goat farming in Indian condition by poor landless and marginal farmers in terms of providing handsome income. Based on feeding habits, feed consumption and maintenance sheep are better than goats. Goats have browsing habits which may lead to damage to the vegetation cover whereas sheep are hardy compared to goats and can thrive on thorny and waste parts of feed and fodders.

The demand of milk is uniformly dispersed throughout the market after the cooperative movement and operation flood in dairy sector. The poultry segment is concentrated in peri-urban region of the city and mainly marketed by aggregators in contract farming with the farmers. About 70% of marketed surplus of milk is handled by the informal sector in India. Besides this, processing of milk is about 20% that needs more investment in coming days due to popularity of value based products among the consumers. No doubt cooperative sector has been playing major role in livestock based product marketing, still the advent of producer companies (eg. Maahi Milk Producer Company) and new entry of ITC in Bihar in milk created motivating environment for other players to invest in this sector. According to Dr Dinesh Bhosale, Regional Sales Director, AB Vista- South Asia, about 50 million broilers are produced per week and out of them 45 million are sold in the live market and rest are processed. It opens a great scope for processing industry.

## ANIMAL FEED SECTOR: A LUCRATIVE AREA FOR INVESTMENT

The area under cultivated fodder production is limited only to 5% of the total cultivable land (Table 9.1). In the states of Haryana, Punjab, Gujarat and some parts of Rajasthan, land use for green fodder production is estimated at 10% or more.

**Table 9.1:** Ingredients used in animal feed

| Cereals and grains | Cakes and meals | Feeds of animal origin | By products used as animal feed |
|---|---|---|---|
| Maize, Sorghum, Bajra | Soybean, Groundnut, Rapeseed (Poultry) and Cotton seed and Copra (Cattle) | Meat meal, fish meal, Bone meal, dicalcium phosphate | Biscuit waste, Chuni, Tulsi seed, Babul seed, Dhaincha seed, Barley waste, Orange peel |

Azolla cultivation is promoted in feeding of livestock by different government schemes of Odisha, mainly through Agriculture Technology Management Agency (ATMA). Fish meal and meat meal were popularly used in poultry feed; and it is now replaced with soybean meal by looking into the preference of former in some areas of demand. Even government plans to import soybean in advance to fulfill the demand in the lean period. Besides that, cattle and poultry feed must be supplemented with vitamins (Vitamin A, $D_3$ and B complex), mineral mixture, probiotics, antibiotic for increased production of milk and meat.

There is a need for restructuring the land use strategy to elevate the overall proportion of cultivable lands for fodder production (Table 9.2). While the cost of feed and fodder combined make about 2/3rd of running expenditure of a dairy farm, the availability of proper feed and fodder at competitive rate has never been matching the demand. There is a big gap between demand and supply of cattle feed where there is opportunity for investment and profit. Moreover, the concept of availability of green fodder and silage as marketable items is yet to grow.

**Table 9.2:** Dairy entrepreneurship development scheme by NABARD

| S.No | Component | Unit Cost | Pattern of Assistance |
|---|---|---|---|
| 1 | Establishment of small dairy units with cross bred cows/ indigenous descript milch cows like Sahiwal, Red Sindhi, Gir, Rathi etc / graded buffaloes upto 10 animals | Rs 5 lakhs for 10 animals unit-minimum unit size is 2 animals with an upper limit of 10 animals | 25% of the outlay (33.33% for SC/ST farmers) as back ended capital subsidy subject to a ceiling of Rs 1.25 lakh for a unit of 10 animals (Rs 1.67 lakh for SC/ST farmers). Maximum permissible capital subsidy is Rs 25000 (Rs 33,300 for SC/ST farmers) for a 2 animal unit. Subsidy shall be restricted on a prorate basis depending on the unit size |

| S.No | Component | Unit Cost | Pattern of Assistance |
|---|---|---|---|
| 2 | Rearing of heifer calves- cross bred, indigenous descript milch breeds of cattle and of graded buffaloes – up to 20 calves | Rs 4.80 lakh for 20 calf unit- minimum unit of 5 calves with an upper limit of 20 calves | 25% of the outlay (33.33 % for SC/ST farmers) as back ended capital subsidy subject to a ceiling of Rs 1.20 lakh for a unit of 20 calves (Rs 1.60 lakh for SC/ST farmers). Maximum permissible capital subsidy is Rs 30,000 (Rs 40,000 for SC/ST farmers) for a 5 calf unit. Subsidy shall be restricted on a prorata basis depending on the unit size |
| 3 | Vericompost (with milch animal unit. To be considered with milch animals and not separately) | Rs 20,000/- | 25% of the outlay (33.33% for SC/ST farmers) as back ended capital subsidy subject to a ceiling of Rs 5,000/-(Rs6700/- for SC/ST farmers) |
| 4 | Purchasing of milking machines/ milktesters/ bulk milk cooling units (upto 2000 lit capacity) | Rs 18 lakh | 25% of the outlay (33.33% for SC/ST farmers) as back ended capital subsidy subject to ceiling of Rs 4.50 lakh (Rs 6.00 lakh for SC/ST farmers |
| 5 | Purchase of dairy processing equipment for manufacture of indigenous milk products | Rs 12 lakh | 25% of the outlay (33.33 % for SC/ST farmers) as back ended capital subsidy subject to a ceiling of Rs 3.00 lakh (Rs 4.00 lakh for SC/ST farmers) |
| 6 | Establishment of dairy product transportation facilities and cold chain | Rs 24 lakh | 25% of the outlay (33.33% for SC/ST farmers) as back ended capital subsidy subject to a ceiling of Rs 6.00 lakh (Rs 8.00 lakh for SC/ST farmers) |
| 7 | Cold storage facilities for milk and milk products | Rs 30 lakh | 25% of the outlay (33.33% for SC/ST farmers) as back ended capital subsidy subject to a ceiling of Rs 7.50 lakh (Rs 10.00 lakh for SC/ST farmers) |
| 8 | Establishment of private veterinary clinic | Rs 2.40 lakh for mobile clinic and Rs 1.80 lakh for stationary clinic | 25% of the outlay (33.33% for SC/ST farmers) as back ended capital subsidy subject to a ceiling of Rs 60,000/- and Rs 45,000/- (Rs 80,000/- and Rs 60,000/- for SC/ST farmers) respectively for mobile and stationary clinic |
| 9 | Dairy marketing outlet / Dairy parlour | Rs 56,000/- | 25% of the outlay (33.33% for SC/ST farmers) as back ended capital subsidy subject to a ceiling of Rs 14,000/- (Rs 18600/- for SC/ST farmers). |

**End Note**

Livestock sector is currently in news for its contribution for ozone layer depletion as methane gas is produced after ruminal fermentation for digestion. At the same time, one of the entrepreneurs of Gujarat exports cow dung in powder form to Dubai where lack of soil affects the agricultural production. Increase value of livestock products has opened the gates for livestock insurance and consultancy in a more competitive way which creates scope for the entrepreneurship in the sector. Livestock production unlike agriculture acts as minibank for entrepreneurs and farmer as it is more liquid in nature. Similarly turn over period for livestock product is very short compared to crop husbandry which generates more income and opportunities.

■■■■

*Chapter 10*

# PROJECT FORMULATION

**Sanjiv Kumar**

*Scientist, ICAR- NAARM, Rajendranagar, Hyderabad*

## INTRODUCTION

A project may be defined as an investment activity where financial resources are utilized to create capital assets with an anticipation of benefits over time. Every project has a pre-defined objective which is to be achieved within a particular time frame. Hence, it has a specific start point and an end point to achieve the objective. It can also be seen as a time bound set of interrelated activities to achieve a particular objcctive. Agri-business projects may include but not limited to establishing a green house, poultry unit, dairy farm, milk processing unit, cold storage unit, food retail enterprise, agricultural credit institution, food processing unit, and warehouse.

All projects are carried out under three constraints, namely **scope/quality**, **cost** and **time**. These are commonly known as **triple constraint** and often represented as a **triangle**. Undue emphasis on any of these constraints will affect the other. Hence, it is a big challenge in front of the project manager to make a balance among these three constraints. For instance, over emphasis on quality will increase the cost as well as the time of completion, whereas more focus on cost may lead to compromise in quality and time. All the projects pass through various phases of project life cycle are as discussed below.

1. **Project identification**: It involves conceiving some new profitable business idea.
2. **Project formulation**: If the identified project *prima facie* seems convincing, a detailed project feasibility study is undertaken, which is dealt comprehensively in later part of the chapter.
3. **Appraisal:** A careful review or appraisal of the project is done to gauge the soundness of the project before implementation.
4. **Implementation:** It is the actual investment phase of the project. The earlier planning phases are actually implemented in this phase.
5. **Monitoring:** It is continuous observation and checking of the project with regard to proper implementation and taking corrective action in case of any deviation from the planned project.

6. **Evaluation:** This can be concurrent or ex-post in nature. It attempts to assess the performance and success of the project.

## Project Formulation Overview

Project formulation involves detailed arrangements for the technical and operational aspects of project implementation such as the costing, financing and scheduling of project activities.

Project formulation begins with drafting of proposal which lays out the preliminary ideas, general description, and detailed technical, financial, marketing and organizational aspects of the proposed project. First preliminary project report is prepared which is the output of feasibility study, and is prepared to support the investment proposal covering various aspects related to technical, financial, economic feasibility. After the preliminary report is approved, detailed project report (DPR) is prepared to firm up the proposal. For companies seeking financial assistance for implementing its business idea, it is essential to prepare and submit a detailed project report. Hence, the whole project formulation can be divided into following phases:

1. Feasibility study:
   a. Technical aspects
   b. Financial aspects
   c. Economic aspects
2. Preliminary project report
3. Detailed project report

## Components of a detailed project report (DPR)

This is the output of project formulation and forms the basis for decision making for the investor or for financial assistance from the financer. It must have wide range of information which will ensure that fund is granted for the project. This is the culmination of all the technical, financial, market, economic analyses pertaining to the project. Along with these, details of the promoter, proposed organizational structure are also incorporated in the DPR.

Following is the list of information which forms the part of DPR. This is not exhaustive but covers majority of the items to be incorporated with varied assortment.

*General description*

The report begins with a general overview of the sector, current status of the industry, future prospects, etc. It also includes the experience, skills, strengths of the promoters, their educational background, earlier experience in related industry or others as entrepreneur or as an employee, any training received in the related field, etc.

*Products and services*

1. Products proposed to be manufactured or produced with detailed description is to be mentioned. Level of quality or unique or proprietary feature of the proposed products is worth to be depicted which impresses the financer as far as the viability of the project is concerned.
2. The price of the proposed products may be arrived at by using any of the pricing method in vogue. Normally for existing product, price is fixed following the existing players' price.

*Marketing plan*

1. Economics- current demand, growth trends, trend in consumer preference.
2. Assessment of existing and potential market demand and supply (both domestic and international) for the product.
3. The proposed product mix and features and benefits of the products.
4. Existing market players in the industry.
5. Proposed market share for the product that can be secured through appropriate marketing strategy.
6. Reasonable selling price of the product.
7. Proposed distribution channels for the sale of the product upto the level of the end customers.
8. Sales forecast for the coming years based on suitable marketing strategy to be adopted.
9. Export potential based on the size of the international market and the price competitiveness of the product.

*Operational plan*

**1. Location**

a. Distance from the nearest railway station, port and airport
b. Distance from the nearest town
c. Locational advantages from raw material availability/end product market point of view
d. Adequacy of housing facilities in the nearby areas
e. Details of other projects coming up in the area
f. Availability of infrastructural facilities like roads, schools, hospitals etc.
g. Distance from the electrical sub-station
h. Availability of water
i. Availability of skilled/unskilled workers
j. NOC from the government for the proposed site
k. Clearances from State pollution control board

**2. Site**

a. Particulars of land; whether free-hold, lease, rent etc. - if land is leased/ rented, terms there of including the period of lease, etc.
b. Adequacy of land area proposed from the angle of future expansions
c. Terrain of the land
d. Land development cost
e. Electrical sub-station and its capacity to meet project's power requirement
f. Whether dedicated sub-station/power line required and costs thereof

**3. Technical aspects**

a. Technology/technical arrangements:
    - Whether technology is proven specifically in Indian conditions
    - Reference plants within and outside the country
    - Performance data of plants based on same technology
    - Latest developments in the area
    - Competing technologies, risk of obsolescence of technology
    - Basis of selection of technology
    - Merits/demerits vis-a-vis other technologies
    - Reputation of licensor of technology
    - Statutory approvals
b. Plant & machinery:
    - List of equipment and machinery
    - Whether the capacity is comparable with international size plant/ is of minimum economic size
    - Competitive quotations for major items of equipment
    - Technical and commercial evaluation of major/critical equipment
    - Reputation of machinery suppliers
    - Easy availability of spares
c. Raw materials and utilities:
    - Details/specifications of all the raw materials
    - Indigenous availability and names of suppliers
    - Availability of imported raw materials, names of international producers
    - Price trends of main raw materials (indigenous and imported) during the past 3 years
    - Risks involved in sustained supply of raw material

d. Manpower:
    - Details of manpower (managerial, supervisory, skilled etc.)
    - Comparison of manpower strength with similar other projects
    - Comparison of salary/wage structure with salaries in same industry/area
    - Availability of manpower
    - Training needs and facilities provided

e. Implementation schedule:

    Timeline for various major activities like:
    - Time required for land acquisition and development work
    - Period required for erection, installation and commissioning
    - Period provided for trial-runs
    - Comparison of implementation schedule with similar projects
    - Experience of project manager.

e. Project monitoring and co-ordination:
    - The strength and experience of the project team.
    - Team members profile and their earlier experience of handling related or other projects.

e. Process- flow chart and description:

    The complete flow diagram of the production process starting from the receiving of the raw material till the product is finished for sale is to be represented.

*Management and organization*

Following should form a part of the management and organization section and has been discussed below:

1. Memorandum and articles of association- whether the proposed activity is included in the objects clause
2. Authorized share capital
3. Borrowing power; whether the company has powers to borrow against suitable securities
4. Directors and board of directors
5. Promoters' agreement
6. Capital structure
7. Promoters' interest- contributions towards equity capital, representation on the board, etc
8. Terms of appointment of directors by each promoter
9. Sharing of responsibilities
10. Procedure for divestment of interest

11. Particulars of other projects promoted/implemented in the past
12. Projects under implementation
13. Report of dun & Bradstreet or similar agencies, if an overseas company
14. Group companies and their operations.

In case of new promoters, following should be included:

1. Family background, qualifications, previous industrial experience
2. Sources of income
3. Details of personal properties/assets of promoters
4. Status of other projects promoted
5. Particulars of associate companies/concerns and their operational results
6. Management and organisational set up
7. Board composition; whether it is broad based
8. Number of full-time directors & their responsibilities
9. Chief executive; qualification, experience, competence
10. Functional executives; qualification, experience, competence
11. Organization - during implementation and later
12. No. of persons recruited for the project and details of project team

*Personal financial statement*

Assets and liabilities and personal net worth of the owners are important components of the detailed project report as it helps in ensuring the financial soundness of the promoters (Table 10.1). This will ultimately help in providing cushion to the project in case of troubled performance of the project in future.

**Table 10.1:** Networth statement of promoter/s

| Liabilities | Assets |
|---|---|
| Current debt (Credit cards, Accounts) | Cash - savings accounts |
| Notes payable | Certificates of deposit |
| Taxes payable | Securities - stocks / bonds / mutual funds |
| Real estate mortgages (describe) | Notes & contracts receivable |
| Other liabilities (specify) | Life insurance (cash surrender value) |
| | Personal property (vehicle, jewelry, etc.) |
| | Retirement funds |
| | Real estate (market value) |
| | Other assets (specify) |
| Total liabilities | Total assets |
| Net worth | |

*Start-up expenses and capitalization*

Entrepreneurs may seek to keep the project cost deliberately low or high, to minimize promoters' contribution or to secure excessive funding (Table 10.2). In assessing the project cost estimates, it is essential to guard against such possibilities. Realistic estimates should be used so as to prevent cost over-runs. While assessing these costs, escalation on account of inflation during the implementation period to be provided for in case of projects with longer implementation schedule exceeding one year. The main components of the project cost and the main issues concerning its appraisal are briefly discussed below.

1. **Land & site development**: Site development expenses including the cost of leveling of land, laying of roads, fencing and gates etc. The property should have clear titles and be mortgagable.
2. **Building**: The cost of proposed building will include the cost of factory, administrative building, civil work for effluent treatment plant, godown, boundary walls, garages, sewers, water tanks, drainage, architect / consultant fees, etc. The costs should be based on the architect's estimates. The rate per unit of construction is to be compared with the cost of identical structures in case of similar projects. This may however vary depending on site conditions and other meteorological and geological factors. While assessing the cost of civil works in respect of projects with a relatively shorter implementation period of upto one year, the current estimates with regard to various building materials as also construction charges could be used. In the case of projects with implementation schedules exceeding one year, the current cost estimates to be suitably inflated by way of appropriate contingency provision for the period of implementation.
3. **Plant & machinery:** Estimates of plant and machinery should be based on quotations obtained from the suppliers. Adequate provision is be made for foundation, installation, spares, piping, insulation needs, etc.
4. **Miscellaneous fixed assets**: It includes furniture, office equipment, tools, vehicles, equipment for distribution of water, steam, power and lighting, laboratory and workshop equipment, effluent disposal plant, fire-fighting equipment, etc. It is necessary to ensure that all necessary/miscellaneous items are included in the cost estimates.
5. **Preliminary & pre-operative expenses**: In the case of new companies, preliminary expenses incurred by the promoters in connection with incorporation/formation of the company and other related legal expenses may be reckoned in the cost estimates. The company would need to incur certain essential expenses during the implementation period, viz. capital issue expenses, expenses for tying up means of financing, interest during construction period, establishment expenses, guarantee commission, commitment charges/upfront fees, start-up expenses, security deposit with electricity boards for power connection, margin for bank guarantee for proposals covered under the export promotion capital goods scheme, etc. A reasonable level could be about 6% of the total start-up cost.

6. **Contingencies:** This is to provide for unforeseen expenses over and above the project cost estimates due to additions or increase in duties, taxes, transportation charges etc. While there can be no hard and fast rule regarding the quantum to be provided, normally it should be 5% of the initial cost.
7. **Margin money for working capital**: Margin money for working capital should invariably be included in the cost estimates taking into account the prevailing norms for inventory, receivables and permissible bank finance. The requirement of stocks of raw materials, consumables, stores, spares, etc. needed for the production cycle, a reasonable level of inventory of finished goods, stock in process, debtors - all at the optimum level of production and running expenses for about a month are to be computed for arriving at the aggregate working capital projections.

**Table 10.2:** Start-up expenses

| **Expenses** | **Amount (Rs)** |
|---|---|
| 0. Buildings/real estate | - |
| 0. Capital equipment | - |
| 0. Location/administration expenses | - |
| 0. Other expenses | - |
| 0. Contingency fund | - |
| 0. Working capital | - |
| **Total Start-up Expenses** | - |

*Means of finance*

The means of finance must conform to a financially sound capital structuring of the company (Table 10.3). Some of the important sources of finance for a project can be:

1. **Share capital**: While a greater amount of share capital would promise well for the financial parameters from the lenders viewpoint, it should, at the same time be ensured that there is an appropriate debt-equity mix so that the company's future recourse to capital market is not found difficult. Also the profit generation should be adequate to service the projected share capital. The right mix of share capital to debt is a prudent balancing act.
2. **Unsecured loans from promoters**: The unsecured loan being raised is normally interest free. In case payment of interest is envisaged on such unsecured loans, it shall be reasonable, normally, less than the rate of interest on the institutional term loan.
3. **Internal accruals:** In case of existing companies, internal accruals invariably form a part of the means of finance.
4. **Promoters' contribution**: A certain percentage of the capital cost of a project to be met by the promoters to ensure that they have reasonable stake in and commitment to the project. Normally, promoters' contributions should

be in the form of equity. Normally promoters' contribution of around 25% of the project cost is considered decent.

5. **Term loans**: This is generally provided by the financial institutions (Banks or non-banks). Depending on the quantum of loan one or more financier may be approached for the term loan. When the term loan is proposed to be sourced from different lenders, it should be ensured that the other expected sources of loans are properly tied up, since any gap in the means of finance will adversely affect the project.
6. **Others**: At times government subsidy/grants are available for certain kinds of projects, depending upon the prevalent government policy. Since these funds are normally realized only towards the end of the project or after commencement of operations, it is up to the promoters to bring in their own funds to bridge the gap till such time that the government funds are received so that there is no delay in the execution of the project.

**Table 10.3:** Sources of finance

| Source | %age Share | Amount (Rs) |
|---|---|---|
| 1. Equity | | |
| 2. Grant/subsidy | | |
| 3. Debt | | |
| Total | | |

**Table 10.4:** Calculation of maximum permissible bank finance

| | Method A | Method B |
|---|---|---|
| 1. Current assets | | |
| 2. Other current liabilities (other than bank borrowings) | | |
| 3. Working capital gap | | |
| 4. Borrower's contribution | 25% of WCG | 25% of CA |
| 5. Maximum permissible bank finance | Lower (WCG-BC) | |

*Abbreviation: WCG- Working Capital Gap; CA- Current Assets.*

*Financial plan*

1. **Profitability projections:** Profitability estimates, projected cash flow and balance sheet are to be made. The projections normally cover the entire duration of the loan. These financial projections should be considered with reference to cost of the project, means of finance thereof and assumptions should be made in respect of schedule of implementation, capacity utilization, cost of production, administrative expenses, sales realization etc. As far as possible the assumptions should be reasonable and realistic and should be relevant to the industry.

**Table 10.5:** Monthly sales statement

| | Apr | May | Jun | Jul | Aug | Sep | Oct | Nov | Dec | Jan | Feb | Mar | Annual Total |
|---|---|---|---|---|---|---|---|---|---|---|---|---|---|
| P-1 Units | | | | | | | | | | | | | |
| S.P. | | | | | | | | | | | | | |
| Total | | | | | | | | | | | | | |
| | | | | | | | | | | | | | |
| P-2 Units | | | | | | | | | | | | | |
| S.P. | | | | | | | | | | | | | |
| Total | | | | | | | | | | | | | |
| | | | | | | | | | | | | | |
| P-3 Units | | | | | | | | | | | | | |
| S.P. | | | | | | | | | | | | | |
| Total | | | | | | | | | | | | | |
| Monthly Total | | | | | | | | | | | | | |

P-1, P-2, P-3 are the units of products-1, 2, 3; S.P is the selling price.

**Table 10.6:** Projected profit and loss account

| Particulars | Year 1 | Year 2 | Year 3 |
|---|---|---|---|
| SALES: | | | |
| GROSS SALES: | | | |
| **ADD: Other Operating Income** | | | |
| NET SALES: | | | |
| Increase in Net Sales (%) | | | |
| COST OF SALES: | | | |
| (i) Raw materials | | | |
| (ii) Other Spares | | | |
| (iii) Power & Fuel | | | |
| (iv) Direct Labour | | | |
| (v) Other Manuf. Expenses | | | |
| (vi) Depreciation | | | |
| (vii) Repairs & Maintenance | | | |
| (viii) Other Manufacturing Expenses | | | |
| SUB TOTAL: | | | |
| **ADD: Opening Stock in Process** | | | |
| DEDUCT: Closing Stock in Process | | | |

| Particulars | Year 1 | Year 2 | Year 3 |
|---|---|---|---|
| COST OF PRODUCTION: | | | |
| COP as % of Gross Sales | | | |
| **ADD: OP Stock of Finished Goods** | | | |
| DEDUCT: CL Stock of Finished Goods | | | |
| COST OF SALES: | | | |
| Cost of Sales as % of Gross Sales | | | |
| Selling, General & Administrative Expenses | | | |
| PROFIT BEFORE INTEREST & TAX (PBIT) | | | |
| PBIT as % of Gross Sales | | | |
| **Interest & Other Financial Charges** | | | |
| Interest & Financial Charges as % of Sales | | | |
| OPERATING PROFIT BEFORE TAX (PBT) | | | |
| PBT as % of Gross Sales | | | |
| **ADD: Other Non-operative Income** | | | |
| (i) Interest & dividend | | | |
| (ii) Exchange Profit/ Export Incentives | | | |
| (iii) Exchange Provision Written Back | | | |
| (iv) Profit on sale of assets | | | |
| (v) Sale of Scrap | | | |
| **SUB-TOTAL (INCOME)** | | | |
| Deduct Other Non-operating Exp. | | | |
| (i) Loss on Investment | | | |
| (iii) Loss on Sale of Fixed Assets | | | |
| (iv) Bad Debts Written-off | | | |
| (v) Miscellaneous Expenses Written off | | | |
| **Sub-total (expenses)** | | | |
| PROFIT BEFORE TAX / LOSS | | | |
| Tax Paid | | | |
| Provision for Taxes | | | |
| NET PROFIT/LOSS (PAT) | | | |
| PAT as % of Gross Sales | | | |
| Equity/Preference Dividend Paid (i) Amount | | | |
| RETAINED PROFIT | | | |
| | | | |

Based on the projections, following ratios should be worked out and analyzed to ensure whether they are satisfactory and conform to norms.

a. Debt-Equity Ratio (DER): Around 1.5:1 is considered to be sound, but this may go up to 2:1 for larger projects.
b. Current ratio (CR): A project is required to have a current ratio of 2:1. It should never go below 1.33:1.
c. Fixed asset coverage ratio (FACR): The fixed asset coverage of at least 2 times seems to be a good indicator.
d. Gross profit
e. Operating profit
f. Net profit
g. Debt service coverage ratio (DSCR): It should be at a comfortable level of at least 2 times over the period of the loan, with the ratio not being below 1 in any given year.
h. Earnings per share
i. Dividend payable
j. Unit cost of production

**Table 10.7:** Statement of financial analysis

| | Year 1 | Year 2 | Year 3 |
|---|---|---|---|
| **A. PROFILE ANALYSIS** | | | |
| Total Assets | | | |
| Total Outside Liabilities | | | |
| Net Worth | | | |
| Net Sales | | | |
| PBDIT | | | |
| Operating Profits (PBIT) | | | |
| Net Profit | | | |
| Gross Cash Accruals | | | |
| Net Working Capital | | | |
| % of NWC to Current Assets | | | |
| **B. LIQUIDITY ANALYSIS** | | | |
| Current Ratio | | | |
| Quick Ratio | | | |
| **C. PROFITABILITY ANALYSIS** | | | |
| PBDIT/Sales (%) | | | |
| PBIT / Net Sales (%) | | | |
| PBT / Net Sales (%) | | | |

| | Year 1 | Year 2 | Year 3 |
|---|---|---|---|
| Net Profit/ Sales (%) | | | |
| Return on Assets (%) | | | |
| Retained Profits / Net Profits (%) | | | |
| Return on Net Worth(%) | | | |
| **D. ACTIVITY ANALYSIS (IN DAYS)** | | | |
| Receivable Turnover | | | |
| Inventory Turnover | | | |
| Accounts Payables Turnover | | | |
| Fixed Assets Turnover Ratio | | | |
| **E. GROWTH RATIOS** | | | |
| Net Sales Growth (%) | | | |
| Net Profit Growth (%) | | | |
| Net Worth Growth (%) | | | |
| **F. LEVERAGE & OTHER RATIOS** | | | |
| TOL / TNW | | | |
| Interest Cover | | | |
| DER | | | |
| FACR | | | |

Some of the key assumptions in this regard are briefly discussed as under:

a. **Capacity utilization:** It generally starts with low level and gradually increases over the years. In most of the projects 60% capacity utilization in the first year is considered good, which is increased to 80-90% in steps of 10-20% on year on year basis. Capacity utilization should never be allowed to reach 100%. A maximum of 80-90% is sufficiently high. Also levels of utilization would vary from product to product. While a higher level of capacity utilization during the optimum year of operation may sometimes be projected in respect of many products, these projections should be compared with the actual capacity utilization being achieved in the industry. In case of products in a highly competitive market environment or in case of a new product, a gradual build-up of capacity utilization is more realistic.

b. **Unit sale price:** The unit sale price may be based on prevailing price in the market. It must accommodate for additional features and services endowed with the product. While determining the sale price, it is necessary to keep in mind important factors such as excise, sales tax etc. The expected changes in these rates in future should also be considered.

c. **Raw material prices:** Besides the existing price of raw materials, it is necessary to check whether they are freely available in the country, their source, etc. in order to determine the likely trend of future price of raw materials. In case of imported raw materials, international price fluctuations should be borne in mind. Raw material prices as prevailing during the past 3-4 years should be studied. Some materials are particularly subject to cyclical changes. These should also be studied closely before assuming a price.

d. **Other operating costs:** Annual increase at the rate of at least 5-10% should be taken for wages and salaries. Repairs and maintenance should be based on the type of industry and the number of working shifts. While these costs may be lower in the first two years due to new equipment, the repairs and maintenance cost should be gradually stepped up in the later years. Advertisement costs may be quite high in case of a new product or in case of a company that does not enjoy any particular brand equity.

e. **Sensitivity analysis:** Since project proposals are estimates of likely culmination of future events, risks are inherent in all proposals. While appraising a project the critical areas of risks should be identified and expressed in terms of quantifiable variables such as price, cost, capacity utilization, etc. Thereafter, the financial projections should be subjected to the tests of sensitivity analysis and the projects viability in such adverse scenarios should be analyzed. In such analysis, the choice of critical assumptions to be varied should be based on the boundary conditions that can be realistically assumed. There can be no fixed norms for the extent to which the critical assumption can be changed. Sensitivity analysis, generally, is carried out by increasing the cost of raw materials by 3- 5% and by reducing the price of the product by 3-5%.

f. **Balance sheet projections**: This should be derived from the cash flow and profitability estimates (Table 10.8). The projected balance sheets at different points of time provide the financial structure of the project as it would evolve, and are extremely helpful in assessing, for example, the adequacy of current ratio, fixed assets coverage ratio, debt equity ratio etc.

**Table 10.8:** Projected balance sheet

| Particulars | Year 1 | Year 2 | Year 3 |
|---|---|---|---|
| **LIABILITIES:** | | | |
| Current liabilities: | | | |
| Short Term Borrowings | | | |
| Sundry Creditors (Trade) | | | |
| Advance payment from customer | | | |
| Provision for Taxation | | | |
| Dividend payable | | | |
| Other Statutory Liabilities (Due within One Year) | | | |
| Installments of Term loans / Debentures / deposits etc. (due within 1 year) | | | |
| Other current liability & provisions (due within 1 year) | | | |

| |
|---|
| Interest accrued bot not due |
| Dues to Directors |
| Security Deposits – Suppliers and Contractors |
| Other Current Liabilities |
| TOTAL CURRENT LIABILITIES |
| |
| **TERM LIABILITIES** |
| Debentures (not maturing within 1 year) |
| Preference Shares (redeemable after 1 year) |
| Term Loans( Excluding installment payable within 1 year) |
| Term Deposits ( repayable after 1 year) |
| Unsecured loans |
| Deferred Sales Tax |
| Deferred Tax Liability |
| TOTAL TERM LIABILITIES |
| |
| Ordinary share capital |
| Preference Share Capital |
| General reserve |
| Capital Reserve |
| Surplus(+) or deficit (-) in P & L Account |
| NET WORTH |
| TOTAL LIABILITIES |
| |
| **ASSETS:** |
| CURRENT ASSETS |
| Cash and Bank Balances |
| Investments (Other than Long Term) |
| Govt. & other securities |
| Fixed deposits with banks |
| RECEIVABLES |
| |
| **INVENTORY** |
| Raw Material |
| Stock in process |
| Finished Goods |
| Goods in Transit |

| Closing Stock of Traded goods |
|---|
| Stores & Spares |
| Advance to suppliers of raw material |
| Other Current Assets: |
| Deposits |
| Duties & Taxes paid in Advance |
| Others |
| Loans and Advances |
| Interest Accrued on Investment & Deposits |
| TOTAL CURRENT ASSETS |
| **FIXED ASSETS** |
| Land |
| Building |
| Plant & Machinery |
| Furniture & Fixtures |
| Other Fixed Assets |
| Capital Works in progress |
| |
| NET BLOCK |
| OTHER NON CURRENT ASSETS |
| Investments in Subsidiary companies |
| Other Investments |
| Advances to suppliers of Capital Goods/Contractors |
| Deferred receivables(maturity > 1 year) |
| Margin money kept with banks. |
| Debtors exceeding 6 months |
| Short Term Deposits |
| Non-consumable stores & spares |
| Other Non-current assets including dues from directors |
| TOTAL OTHER NON CURRENT ASSETS |
| Intangible Assets (patents, goodwill, prelim. expenses) |
| TOTAL ASSETS |
| |
| **Total Liabilities - Total Assets** |

2. **Cash flow / funds flow projections:** The cash flow statements help in studying the sources and uses of funds both during implementation and operation (Table 10.9 and 10.10). Not only should the projections show cash accruals, but

there should be provision for sufficient cash availability to meet the operational requirements of the project as well as the repayment of the loan. The amortization schedule of the term loan should be fixed on the basis of the cash flow projections. The financial soundness of a project is determined by the surplus which it can generate during the period of the loan. Hence the appraisal has to ensure that the surplus shall be available not only to meet the loan commitments on time but also to service the risk capital to a reasonable extent.

**Table 10.9:** Funds flow statement

| **Particulars** | **Year 1** | **Year 2** | **Year 3** |
|---|---|---|---|
| Profit After Tax | | | |
| Depreciation | | | |
| Dividends | | | |
| Funds From Operations | | | |
| Long Term Sources: | | | |
| Change in Capital | | | |
| Net Change in Reserve | | | |
| Change in Term Loans | | | |
| Total [Source(+) / Deficit(-)] | | | |
| Long Term Uses: | | | |
| Net Change in Fixed Assets | | | |
| Change in Other Non-Current Assets (including Investments) | | | |
| Change in Intangibles | | | |
| Contribution to Working Capital | | | |
| Short Term Uses | | | |
| Change in Net Working Assets | | | |
| Change in Other Current Assets | | | |
| Short Term Sources | | | |
| Change in Other Current Liabilities | | | |
| Change in Bank Borrowings/Loans | | | |
| Net Deficit/ Surplus in Short Term Sources | | | |
| | | | |
| Net Movement in Liquid Assets | | | |
| Change in cash | | | |
| Change in Marketable Investments | | | |
| **Net Movement in Liquid Assets** | | | |

**Table 10.10:** Cash flow statement

| **Particulars** | **Year 1** | **Year 2** | **Year 3** |
|---|---|---|---|
| Operating Sector | | | |
| **Sales** | | | |
| **Debtors (Trade)** | | | |
| Cash From Sales | | | |
| Costs- Interest & Finance Charges | | | |
| **Expenses on Stocks Purchases** | | | |
| **Trade Creditors** | | | |
| **Manufacturing Expenses** | | | |
| Cash Cost of Sales | | | |
| **Expenses for Increase /Decrease in Stocks** | | | |
| Cash from Asset Conversion Cycle | | | |
| **Selling, Gen. & Adm. Expenses** | | | |
| **Advance Payments** | | | |
| **Advances Received** | | | |
| **Taxation** | | | |
| **Dividends** | | | |
| Cash From Operations | | | |
| **Other Current Assets** | | | |
| **Other Current Liabilities** | | | |
| **Other Income/Expenses(Net)** | | | |
| Net Cash From Operations | | | |
| Investment Sector | | | |
| **Capital Expenditure** | | | |
| **Investment in Group Companies** | | | |
| **Intangible/Other Term Assets** | | | |
| Cash Before Funding | | | |
| Financing Sector | | | |
| **Dues to Banks** | | | |
| **Short Term Debts** | | | |
| **Term Debts** | | | |
| **Equity** | | | |
| **Other Loans & Reserves** | | | |
| Total | | | |
| Movement in Cash Assets | | | |
| **Cash & Bank Balance** | | | |
| **Investments (Other Than Long Term)** | | | |
| Movement in Cash Assets | | | |

3. **Break Even Point (BEP):** The main purpose of the BEP analysis is to determine the lowest production and sales levels at which the project will cover all its costs (Table 10.11). The lower the BEP, the better it is for the project's success.

**Table 10.11:** Break-even analysis

| **Cost Description** |
|---|
| **Variable Costs** |
| Inventory |
| Raw Materials |
| Direct Labor |
| |
| **Fixed Costs** |
| Salaries |
| Supplies |
| Repairs & maintenance |
| Advertising |
| Car, delivery and travel |
| Accounting and legal |
| Rent |
| Telephone |
| Utilities |
| Insurance |
| Taxes (Real estate, etc.) |
| Interest |
| Depreciation |
| Other (specify) |
| Miscellaneous expenses |
| |
| Total Fixed Costs |
| |
| Total Variable Costs |
| |
| |
| Breakeven Sales level = |

The following formula should be used to estimate the Break Even sales level:

$$\text{Breat Even Sales Level} = \frac{\text{Total Fixed Cost}}{\text{Selling Price} - \text{Variable cost per unit}}$$

4. **Internal Rate of Return (IRR):** It is, based on the discounted cash flow method. IRR is the rate of discount that equates the present value of future cash benefits (cash inflows) to the present value of capital cost over the economic life of the project (cash outflows), normally taken at 15 years. In certain industries where technological obsolescence is very high, the project life may be calculated for a lesser period, say 10 to 12 years. Both pre-tax and post-tax IRR are to be calculated. A post-tax IRR of 15% and at the same time ensuring a margin of 2% above the weighted average cost of capital (post-tax) for the project may be considered reasonable. IRR is to be linked to interest rate in capital formation.

*Case*

**A hypothetical detailed project report (DPR) for establishing a dairy farm of 15 Holstein Friesian Cows**

1. **General description**: Dairying is an important source of subsidiary income for farmers. Various products prepared from cow milk form a major food component. Milk products like ghee, butter, cheese, paneer, milk-powder etc. being very nutritious are very popular in Indian foods and have a great demand in the market. So, there is a great opportunity in dairy farming. Dung has its utility as manure and provides a good source of organic matter for improving soil fertility and crop yields. The biogas from the dung is one of the natural source of energy used as fuel for domestic purposes as also for running engines for drawing water from well. The surplus fodder and agricultural by-products can be gainfully utilized for feeding the animals. Since crop agriculture is mostly seasonal, there is a possibility of finding employment throughout the year for many persons through dairy farming. Thus, dairy also provides employment throughout the year.
2. **General background of the promoter:** Mr. Kailash is a farmer having 10 hectares of land. He is a resident of Rampur village in Fatuah block of Patna district in Bihar. He is in farming since long. He is a graduate and has attended a certificate course in dairy farming from Patna Veterinary College.
3. **Location and land details:** The project land is situated at road side of the village and is surrounded by the green trees making the site cool and pleasant. Land is already at a high level making it suitable for farm. Fencing work has to be done on the land.
4. **Infrastructure:** The cattle shed of concrete flooring along with surrounding wall with tin roofing is proposed to be built. The tin roofing will be supported by steel channels and angles. The whole cattle shed will be fenced by steel net. A storage godown will have to be constructed to store cattle feed and fodder. The cattle shed would provide sufficient light and air to the

cattle. The cattle shed would be sufficient for keeping the proposed cattle. The project land is ideal and sufficient for the construction of this infrastructure.

5. **Animals**

*Proposed features*

The proposed features have been discussed below:

1. 15 high breed HF cows.
2. Milk production capacity up to 16 liters per day.
3. Lactation days up to 300 days in a year.

It is proposed to purchase 15 cross-bred HF cows from trader from Bangalore having production capacity of 16 litres per day. The cow normally gives milk for 300 days in a year. Each cow gives birth to a calf every year. A female calf becomes productive after 3 years from its birth. A male calf can be sold in the market for agricultural purpose after feeding it for one year. Healthy, high yielding animals will be purchased with help of veterinary doctor and these animals will be identified by using RFID tag. New purchased animals will be kept under observation for at least two weeks and then will be allowed to mix with the herd.

*Man power*

**Requirement**: One person per 10 cows with salary Rs. 3500/- per month. So, 2 persons will be required. Labour is easily available in the village. They will be engaged on full time basis.

*Procurement of cattle feed*

1. **Requirements:** Sudha compound feed, dry fodder, green fodder will be fed to the animals in appropriate proportion and quantity.
2. **Availability:** Sudha Daan is available from milk co-operative society. Dry and green fodder are available from own farm and will also be purchased from other farmers in the village. Dry fodder for 240 days is required to be kept in stock. It is available at low price in season so it would be profitable to purchase in bulk and store it for the off season. Stock for 30 days consumption of Sudha Daan is required.

*Market feasibility*

Animals will be milked twice a day and the milk will be sold to the Sudha Dairy co-operative. The milk will also be sold to the local market of fatuah, especially sweet shops. Demand in the area is very high. So, there will not be any problem of sale of milk

*Financial feasibility*

The financial viability of the project is determined on the basis of the financial calculations (Table 10.12 and 10.13) along with assumptions which are as follows:

1. Cost of project and means of finance
2. Projected balance sheet
3. Projected profit & loss account (Table 10.14)
4. Cash and fund flow statement (Table 10.15)
5. Repayment schedule and interest calculation
6. Ratio analysis (Table 10.17)

**Table 10.12:** Cost of project

| **Cost of Project** | **Units** | **Rate** | **TOTAL** |
|---|---|---|---|
| Cow/Buffalo (inclusive of transportation & insurance) | 15 | 50,000 | 7,50,000 |
| Shed & Godown Construction | | 4,50,000 | 4,50,000 |
| Grand Total | | | **12,00,000** |

**Table 10.13:** Means of finance

| **Means of Finance** | **%** | **Proposed** |
|---|---|---|
| Borrower's Margin | 20 | 2,40,000 |
| Term Loan from State Bank of India | 80 | 9,60,000 |
| Grand Total | | **12,00,000** |

**Table 10.14:** Projected profit and loss statement

| | **For the Year Ending (Rs. in Lakhs)** | | | | |
|---|---|---|---|---|---|
| | **2016** | **2017** | **2018** | **2019** | **2020** |
| REVENUE | | | | | |
| Sale of Cow Milk | 19.89 | 21.88 | 24.06 | 26.47 | 29.12 |
| Sale of Cow Dung | 0.41 | 0.45 | 0.49 | 0.54 | 0.59 |
| Sale of Empty Bags | 0.39 | 0.41 | 0.43 | 0.45 | 0.48 |
| TOTAL REVENUE | 20.69 | 22.73 | 24.99 | 27.46 | 30.19 |
| COST | | | | | |
| Cost of Feeding Animals | 12.75 | 14.02 | 15.42 | 16.97 | 18.66 |
| Salary / Wages | 0.84 | 0.92 | 1.02 | 1.12 | 1.23 |
| Insurance | 0.45 | 0.47 | 0.50 | 0.52 | 0.55 |
| Medicine | 0.30 | 0.32 | 0.33 | 0.35 | 0.36 |
| Electricity Expense (Rs.1,500 per month increased @ 10% p.a.) | 0.18 | 0.20 | 0.22 | 0.24 | 0.26 |
| Other Operational cost (@ 1% of Revenue) | 0.21 | 0.23 | 0.25 | 0.27 | 0.30 |
| Depreciation | 0.45 | 0.41 | 0.36 | 0.33 | 0.30 |
| Interest on Term Loan | 1.11 | 0.93 | 0.71 | 0.45 | 0.17 |
| Interest on working capital loan @ 12.95 % p.a. | - | - | - | - | - |

| | For the Year Ending (Rs. in Lakhs) | | | | |
|---|---|---|---|---|---|
| | **2016** | **2017** | **2018** | **2019** | **2020** |
| TOTAL COST | **16.28** | **17.50** | **18.81** | **20.25** | **21.83** |
| PROFIT BEFORE TAX | 4.40 | 5.24 | 6.18 | 7.22 | 8.36 |
| Tax | 1.32 | 1.57 | 1.85 | 2.16 | 2.51 |
| NET PROFIT AFTER TAX | **3.08** | **3.67** | **4.33** | **5.05** | **5.85** |
| Add: Depreciation | 0.45 | 0.41 | 0.36 | 0.33 | 0.30 |
| Profit After Tax (PAT) | **3.53** | **4.07** | **4.69** | **5.38** | **6.14** |

**Table 10.15:** Projected cash flow statement

| **Heads** | **2016** | **2017** | **2018** | **2019** | **2020** |
|---|---|---|---|---|---|
| Opening Cash & Bank Balance | - | 0.90 | 0.42 | 0.38 | 0.56 |
| | | | | | |
| Cash Flow From Operations | | | | | |
| **Net Cash profit** | 3.53 | 4.07 | 4.69 | 5.38 | 6.14 |
| Total Cash Inflow: | **3.53** | **4.07** | **4.69** | **5.38** | **6.14** |
| Repayment of Principal | 0.83 | 1.82 | 2.05 | 2.30 | 2.59 |
| Cash Withdrawing | 1.80 | 1.98 | 2.18 | 2.40 | 2.64 |
| Reinvestment in Business | | 0.75 | 0.50 | 0.50 | 0.50 |
| Total Cash Outflow: | **2.63** | **4.55** | **4.73** | **5.20** | **5.72** |
| Closing Cash & Bank Balance | **0.90** | **0.42** | **0.38** | **0.56** | **0.98** |

**Table 10.16:** Projected balance sheet

| **Heads** | As on 31st March | | | | |
|---|---|---|---|---|---|
| | **2016** | **2017** | **2018** | **2019** | **2020** |
| CAPITAL & LIABILITIES | | | | | |
| CAPITAL | | | | | |
| Owner's Capital a/c | 3.68 | 5.37 | 7.52 | 10.17 | 13.39 |
| BORROWINGS | | | | | |
| Bank Loan- Term Loan | 8.77 | 6.94 | 4.89 | 2.59 | (0.00) |
| TOTAL LIABILITIES | **12.45** | **12.31** | **12.41** | **12.76** | **13.39** |
| | | | | | |
| ASSETS | | | | | |
| FIXED ASSETS | 11.55 | 11.15 | 10.78 | 10.45 | 10.16 |
| INVESTMENTS | | | | | |
| Reinvestment in Business Assets | - | 0.75 | 1.25 | 1.75 | 2.25 |
| CURRENT ASSETS | | | | | |
| Cash & Banks: | 0.90 | 0.42 | 0.38 | 0.56 | 0.98 |
| TOTAL ASSETS | **12.45** | **12.31** | **12.41** | **12.76** | **13.39** |

**Table 10.17:** Ratio analysis

| Heads | 2016 | 2017 | 2018 | 2019 | 2020 |
|---|---|---|---|---|---|
| Net Profit Ratio | | | | | |
| Net profit | 3.08 | 3.67 | 4.33 | 5.05 | 5.85 |
| Net sales | 20.69 | 22.73 | 24.99 | 27.46 | 30.19 |
| | 14.90% | 16.13% | 17.31% | 18.39% | 19.38% |
| **Interest coverage ratio** | | | | | |
| Earnings before interest & tax | 5.51 | 6.17 | 6.89 | 7.67 | 8.52 |
| Interest | 1.11 | 0.93 | 0.71 | 0.45 | 0.17 |
| | 4.98 | 6.61 | 9.73 | 16.91 | 50.81 |
| **Debt service coverage ratio** | | | | | |
| Cash profit after tax + interest | 4.64 | 5.00 | 5.40 | 5.83 | 6.31 |
| Installment | 1.94 | 2.76 | 2.76 | 2.76 | 2.76 |
| | 2.39 | 1.82 | 2.00 | 2.12 | 2.29 |
| **Debt equity ratio** | | | | | |
| Debt | 8.77 | 6.94 | 4.89 | 2.59 | (0.00) |
| Equity | 3.68 | 5.37 | 7.52 | 10.17 | 13.39 |
| | 2.38 | 1.29 | 0.65 | 0.25 | 0.00 |
| **Fixed assets turnover ratio** | | | | | |
| Revenue | 20.69 | 22.73 | 24.99 | 27.46 | 30.19 |
| Fixed assets | 11.55 | 11.90 | 12.03 | 12.20 | 12.41 |
| | 1.79 | 1.91 | 2.08 | 2.25 | 2.43 |
| **Asset coverage ratio** | | | | | |
| Total fixed assets + stock + debtors | 11.55 | 11.90 | 12.03 | 12.20 | 12.41 |
| Outstanding term loan + o/s w.c. Loan | 8.77 | 6.94 | 4.89 | 2.59 | (0.00) |
| | **1.32** | **1.71** | **2.46** | **4.71** | **0.00** |

Miscellaneous tables used for deriving values in above tables (Annexures)

**Table 10.18:** Calculation of depreciation (Rs. in Lakhs)

| ASSETS | Cost | Rate | 2016 | 2017 | 2018 | 2019 | 2020 |
|---|---|---|---|---|---|---|---|
| Cow/Buffalo | 7.50 | 0% | | | | | |
| Shed & Godown Construction | 4.50 | 10% | 0.45 | 0.41 | 0.36 | 0.33 | 0.30 |
| Total Depreciation for the Year | | | **0.45** | **0.41** | **0.36** | **0.33** | **0.30** |
| CLOSING BALANCE OF ASSETS | | | **2016** | **2017** | **2018** | **2019** | **2020** |
| Cow/Buffalo | | | 7.50 | 7.50 | 7.50 | 7.50 | 7.50 |
| Shed & Godown Construction | | | 4.05 | 3.65 | 3.28 | 2.95 | 2.66 |
| **TOTAL VALUE OF ASSETS** | | | **11.55** | **11.15** | **10.78** | **10.45** | **10.16** |

**Table 10.19:** Revenue from sale of milk (Rs. in Lakhs)

| Heads | 2016 | 2017 | 2018 | 2019 | 2020 |
|---|---|---|---|---|---|
| Milk Production | 0.64 | 0.64 | 0.64 | 0.64 | 0.64 |
| Rate of Cow Milk | 27.00 | 29.70 | 32.67 | 35.94 | 39.53 |
| Revenue From Sale of Cows Milk | 17.29 | 19.02 | 20.93 | 23.02 | 25.32 |
| Bonus at 15% (Given by Sudha Cooperative) | 2.59 | 2.85 | 3.14 | 3.45 | 3.80 |
| Total Revenue | **19.89** | **21.88** | **24.06** | **26.47** | **29.12** |

**Table 10.20:** Details of milk production

| | |
|---|---|
| Lactation period per annum (days) | 305 |
| Dry period per annum (days) | 60 |
| No of Cow | 15 |
| Avg. Milk Per Day per cow (litres) | 14 |
| Total Daily Production | 210 |
| **Total annual production (litres)** | **64,050** |

**Table 10.21:** Revenue from sale of cow dung (Rs. in Lakhs)

| Heads | 2016 | 2017 | 2018 | 2019 | 2020 |
|---|---|---|---|---|---|
| Number of Cows | 15 | 15 | 15 | 15 | 15 |
| No. of trollies of cow dung per animal p.a. | 2 | 2 | 2 | 2 | 2 |
| Sale Price of cow dung per trolley | 1,800 | 1,980 | 2,178 | 2,396 | 2,635 |
| Total Revenue from sale of Cow Dung | 0.41 | 0.45 | 0.49 | 0.54 | 0.59 |
| Normally one trolley of dung is produced by each animal per annum | | | | | |

**Table 10.22:** Revenue from sale of empty bags (Rs. in Lakhs)

| Heads | 2016 | 2017 | 2018 | 2019 | 2020 |
|---|---|---|---|---|---|
| No of Empty Bags of Sudha daan | 785 | 785 | 785 | 785 | 785 |
| Sale price of Empty Bags | 50 | 53 | 55 | 58 | 61 |
| Total Revenue from sale of Empty Bags | 0.39 | 0.41 | 0.43 | 0.45 | 0.48 |
| Empty bags of Daan are normally sold in local market @ 50 per bag | | | | | |
| Sale Price of Empty Bags increases @ 5% p.a. | | | | | |

**Table 10.23:** Calculation of empty bags of daan consumed by cow

| Heads | No. of animals | Qty per day (Kg) | days per annum | Total Annual Consumption |
|---|---|---|---|---|
| Total Daan consumption (Kg) | 15 | 12 | 305 | 54 900 |
| No of Bags required per annum | | | | 785 |
| One Bag Contains 70 kg of Sudha daan | | | | |

**Table 10.24:** Annual Cost of Feeding (Rs. in Lakhs)

| **Heads** | **2016** | **2017** | **2018** | **2019** | **2020** |
|---|---|---|---|---|---|
| No. of cow | 15 | 15 | 15 | 15 | 15 |
| Annual feeding cost for each cow | 0.85 | 0.93 | 1.03 | 1.13 | 1.24 |
| Total Cost of Feeding | **12.75** | **14.02** | **15.42** | **16.97** | **18.66** |

**Table 10.25:** Cost of feeding each cow (Rs. in Lakhs)

| **Head** | **Qty. (Kg)** | **Rate/Kg** | **Total (Kg)** | **Annual Cost (Rs)** |
|---|---|---|---|---|
| During Lactation (305 Days) | | | | |
| Dry Fodder | 15 | 4.00 | 60 | |
| Green Fodder | 15 | 4.00 | 60 | |
| Sudha daan | 5 | 15.00 | 75 | |
| Other concentrate | 4 | 15.00 | 60 | |
| Total consumption per Cow per day | | | 255 | 0.78 |
| During Dry Days (60 Days) | | | | |
| Dry Fodder | 15 | 4.00 | 60 | |
| Green Fodder | 15 | 4.00 | 60 | |
| Total consumption per Cow per day | | | 120 | 0.07 |
| Grand Total (Rs. lakhs) | | | | 0.85 |

**Table 10.26:** Labour cost (Rs. in Lakhs)

| **Heads** | **2016** | **2017** | **2018** | **2019** | **2020** |
|---|---|---|---|---|---|
| No. of Animals | 15 | 15 | 15 | 15 | 15 |
| No. of Staff (1 person per 10 animals) | 2 | 2 | 2 | 2 | 2 |
| Salary per Annum per Person (Monthly 3500) | 0.42 | 0.46 | 0.51 | 0.56 | 0.61 |
| Annual Labour cost | **0.84** | **0.92** | **1.02** | **1.12** | **1.23** |
| Salary per person per month is Rs. 3500/- | | | | | |
| Salary is increased @10% per annum | | | | | |

**Table 10.27:** Insurance cost (Rs. in Lakhs)

| **Heads** | **2016** | **2017** | **2018** | **2019** | **2020** |
|---|---|---|---|---|---|
| Value of Cows | 7.50 | 7.50 | 7.50 | 7.50 | 7.50 |
| Insurance cost per animal @ | 6.00% | 6.30% | 6.62% | 6.95% | 7.29% |
| Annual Insurance Cost | 0.45 | 0.47 | 0.50 | 0.52 | 0.55 |
| Insurance cost is 6% of the purchase value of productive animal. | | | | | |

**Table 10.28:** Medicines cost (Rs. in Lakhs)

| Heads | 2016 | 2017 | 2018 | 2019 | 2020 |
|---|---|---|---|---|---|
| Total no. of animals | 15 | 15 | 15 | 15 | 15 |
| Cost of medicines per animal per annum | 2 000 | 2 100 | 2 205 | 2 315 | 2 431 |
| Annual Medicine Cost | 0.30 | 0.32 | 0.33 | 0.35 | 0.36 |
| Medical cost per animal per annum is Rs. 2000/- | | | | | |
| Medical cost is increased @5% per annum. | | | | | |

■■■■

*Chapter 11*

# AGRI ENTREPRENEURSHIP SUPPORT SYSTEM

**Baljeet Singh**

*Doctoral Scholar, GH Patel Post Graduate Institute of Management, Sardar Patel University, V.V. Nagar, Gujarat*

## INTRODUCTION TO PROBLEMS FACED BY AGRI START- UPS

An entrepreneur faces large number of hurdles ranges from funds to scientific expertise. Absence of a strong and structured support system leads to high failure of start-up agri-enterprises. It is evident that agri starts-ups are challenged by a huge number of obstacles like land, funds, scientific expertise, testing and trial facilities, managerial acumen, subjective mentorship, office space, pilot production, design validation, and test marketing etc.

### Problem of Funds

An agri start- up may just have an innovative idea promising to offer solution to an orbit shifting challenge but lacks funds. In such a situation, letting the idea die its own death could be a viable but most undesired option. The problem is further aggravated by the fact that banks who are otherwise financers to big projects, do not fund an idea, may it be the most innovative one. In this scenario, the start-up innovator/entrepreneur is in a great dilemma whether to pursue his idea or drop it. Venture Capitalists (VCs) and Angel Investors are a new breed of financers willing to fund innovative ideas or start-up entrepreneurs. Recently, agri-business start-ups have attracted the attention of many investors. They generally fund the start- up sufficiently, without too many restrictions, provided the idea is viable but ask for an equity stake in the company. A few examples of venture capitalists are given below.

1. **Acumen fund** (*www.acumen.org*): Acumen raises charitable donations to invest in companies, leaders, and ideas that are changing the way the world tackles poverty. Started as an idea, now investments from acumen have impacted over 100 million people around the world. They invest with start ups in sectors like agriculture, education, energy, health, housing and water. Recently, Acumen has invested in **Basix Krishi** which is increasing productivity through agricultural extension services to small holder farmers

such as soil testing, seed treatment, vermi composting, pest management and good harvest practices for crops, livestock vaccination, para vet services, fodder demonstration and milk market linkages for dairy farmers. Another investee company of acumen is **Global Easy Water Products (GEWP) Private Limited** (*http://gewp-india.com*) providing affordable drip irrigation solutions to small holder farmers. GEWP's strategy is to leverage rural market forces. The company sources from local manufacturers, uses a network of independent retail dealers, and builds expertise in local technicians to ensure that final product reaches the smallholder farmer at affordable rates.

2. **Omnivore partners** (*www.omnivore.vc*): A venture fund investing in early stage agriculture and food technology companies in India. It supports entrepreneurs who are developing solutions to improve farm productivity, increase agricultural sustainability, modernize agribusiness supply chains and promote farm sourced food products. It seeks to transform Indian agriculture through technology innovation, impacting the lives of farmers and rural communities. For example Omnivore partners have invested in Skymet Weather Services (*www.skymet.net*) which is a leading provider of risk solutions for weather and agriculture (please see box-1), Khedut (Gujarati equivalent of farmer), Agro-engineering (*www.khedutagro.com*) which is one of the leading manufacturer of automatic seed drill and innovative planting machinery, Frontal Rain Technologies (*www.frontalrain.com*), offering affordable cloud based software for food and agri-business companies, Arohan Foods (*www.arohanfoods.com*) -a pork processor sourcing from smallholder farmers across Northeast India and retailing premium pork products nationwide, Barrix Agro Sciences (*www.barrix.in*) which develops integrated pest management (IPM) that provides farmers a highly effective crop protection solutions while reducing their dependence on agro chemicals, Stellapps Technologies (*www.stellapps.com*), a dairy technology solutions provider building automation tools, integrated with cloud, mobility and data analytics for dairy farms, cooperatives and private dairies, Eruvaka Technologies (*www.eruvaka.com*) manufactures on farm diagnostic equipment for aquaculture which integrates sensors with decision tools for affordable monitoring and automation, Mitra (*www.mitraweb.in*) - a leading manufacturer of specialized machinery for horticulture developing automated solutions for cultivation of fruits and vegetables and Y- Cook (*www.ycook.in*), a leading player in retort food processing technology producing pre-cooked vegetables, fruits and legumes that are ambient stable and preservative free.

**Box 1**

***Skymet, funded by Omnivore Partners***

Skymet leads weather forecast by innovation and has a multitude of forecast solutions catering to various industry verticals. Skymet is the only private weather forecasting agency in India that runs its own numeric weather prediction models. It has developed (3 day) precipitation at village-level granularity, 24/48 hour weather forecast in 15 minute increments, medium term (15 day) daily precipitation forecast, seasonal and monsoon forecasts. Key sectors include power companies, media conglomerates, farmer information services, agricultural input producers, logistics operators.

3. **Aavishkaar venture fund** (*www.aavishkaar.in*): It was founded in 2001, Aavishkaar identifies capable entrepreneurs, provides them with capital, supplements with a nurturing environment and helps build sustainable enterprises. Aavishkaar has invested in enterprises that gainfully engage rural and economically week populations either as producers, users or owners to deliver commercial returns. For example Aavishkar has invested in Osam Dairy (*www.osamdairy.com*), a Ranchi based innovative dairy farm run on scientific lines (a part of HR food processing private limited) and also in INI Farms (*www.inifarms.com*), modern horticulture farms, exporting mainly banana and pomegranate, following all advanced standards like good agricultural practices (GAP). INI farms works with land owners but has full control on production from land preparation to final harvesting. It works with local farmers and assists them by providing inputs, knowledge, technology and market access (contract farming). They also provide market access to other farmers producing good quality fruits. Milk Mantra is another investee company of Aavishkar. Milk Mantra (*www.milkmantra.com*) offers natural, healthier milk products using most innovative and advanced state of the art technologies. Its **Ethical Milk Sourcing Programme** ensures that not only does the farmer get the right price each and every time, but also, its network farmers are able to access a variety of extension services aimed at bringing in quality and clean milk production. Via this programme it aims to bring in overall dairy development in Orissa and its neighboring states.

   Shree Kamdhenu Electronic Private Limited (SKEPL) (*www.akashganga.in*) based out of Anand (Gujarat) is an investee company of Aavishkar. It was founded in 1996. SKEPL is an ISO 9001:2008 certified company. It conceptualized the need of **automatic milk collection system** (AMCS) in early 1990 for ensuring transparency, mutual faith and error free operations of milk collection by integrating electronic weighing scales with quality milk testing equipment and data processor/computer.

4. **Small enterprise assistance funds (SEAF):** SEAF established India Agri-business International Fund (SIAIF) in 2010 to target promising SMEs across thirty-three identified sub-sectors of India's agri-business value chain.

At the time of the fund's establishment, 40% of SEAF investments worldwide had been in the food and agri-business sector and SEAF had worked for 6 years with Kotak Mahindra Bank in India on the India growth fund. Bolstered by the local team of experts in agri-business finance, SEAF intends applying our global agri-business experience, along with deep local sector knowledge, to generate substantial sustainable growth in dynamic Indian agri-businesses.

Examples of Angel Investors include Rajasthan Angel Investors Network (RAIN), Gujarat Angels, and Mumbai Angels etc. Angel investors mostly look up for ideas that aim for profit, commercially viable ventures which have potential for high growth/scale, feasibility has been established and founders are committed full time and preferably with a crisp business plan.

An example of an investee company of Angel investor is E-farm, a portfolio company of Mumbai Angels. E-farm is a young social enterprise firm based in Chennai. It offers end to end agri supply chain platform providing a combination of technology solutions and on ground distribution mechanism to enable farmers reach end markets in an effective manner. It ties in farmers, intermediaries, logistics providers, distributors, small time retailers and local road side vendor into a single chain backed up by best of breed information system to deliver fresh, clean, reasonably priced farm produce. Please see *http://www.efarm.in.*

For biotech based innovations, Biotechnology Industry Research Assistance Program (BIRAP) of Biotechnology Industry Research Assistance Council (BIRAC) of Government of India provides limited funding assistance to individuals, companies for refining, validating (proof of concept) of their innovative biotech based ideas. A necessary condition in many of these schemes is that the beneficiary start-up/company should have been incubated in a concerned business incubator. BIRAC has been set up as a section- 8 not for profit company as an interface agency of Department of Biotechnology (DBT) to serve a single window for emerging biotech companies. Examples of BIRAC schemes include:

1. **Biotechnology Industry Partnership Program (BIPP):** Grant in aid type of financial assistance with a matching contribution by the company. Any single or a consortium of Indian companies having Department of Scientific and Industrial Research (DSIR) recognized in-house R&D unit (in case of consortia, at least the main applicant company should have DSIR recognized in-house R&D lab) alone or in partnership with another company/institute/ organization/ university can apply for this grant. Selection for grants is based on quality of proposals submitted.
2. **Small Business Innovation Research Initiative (SBIRI):** It was launched in 2005 to boost public private partnership efforts in the country with an objective to support early stage pre-proof of concept research in Biotechnology by industry, to support new indigenous technologies particularly those related to societal needs, food and nutrition, healthcare, agriculture and other sectors and nurture and mentor innovative emerging

technologies/entrepreneurs to assist new enterprises to forge appropriate linkages with academia and government.

In **phase-I** funding support (grant) to the tune of Rs. 20 lakhs (project cost Rs. 25 lakhs) or 50% of project cost (project cost between Rs. 25 lakhs and Rs. 100 lakhs) is provided. If the project cost is above Rs. 100 lakhs then in addition to above, interest free loan on 50% of the amount exceeding Rs. 100 lakhs is extended. In **phase-II**, soft loan upto Rs. 10 crores is provided for a project as per its requirement. Soft loan upto Rs. 100 lakhs carry a simple interest of 1% while the interest rate is 2% (simple interest) on the amount of loan beyond Rs. 100 lakhs. The role of public R&D institution at this stage too is critical, as many of the projects continue to require technical support from the public funded R&D institutions. The partner in the public institution at this stage gets the R&D support as grant. To get benefit of this scheme, a company can apply alone or jointly with national R&D organization/ institution. Collaborative projects of common interest to the concerned sector/area proposed by a group of industries/users, national research organizations etc. are also eligible to apply under this scheme.

3. **Biotechnology Ignition Grants (BIG):** Biotechnology start- ups/ entrepreneurs having a registered company incorporated on or after 1st January, 2012 or academicians, scientists, researchers, Ph.Ds, medical degree holders, biomedical engineering graduates incubated in a Technology Business Incubator (TBI) are eligible to apply for funding support under this scheme.
4. **Contract Research Scheme (CRS):** Indian company (ies) registered under the Indian companies act, 1956 with at least 51% Indian shareholding having DSIR recognized in house R&D unit or public/private universities and research institutions having well established support system for basic and/or applied research with registration/recognition from a statutory body are eligible to apply for funding support under this scheme.
5. **Micro, Small and Medium Enterprises Schemes:** Support for managerial and entrepreneurial development of SMEs through Incubators is a scheme run by Ministry of MSME being implemented through incubators where innovative ideas from start- up innovators are awarded around Rs. 6 lakhs per idea. The aim is to award 10 ideas per incubator with a disbursement target of Rs. 62 lakhs per annum. An initial screening of the ideas is done at incubator level while final one is held at MSME office, New Delhi. The disbursement of funds is through the incubators and not directly to the innovator. Hence, any innovator willing to take the benefit of the scheme has to apply through concerned incubator only.
6. **Small Industries Service Institute (SISIs):** They are located in different parts of the country (one in each state) are administered by Small Industries Development Organization (SIDO). The following are the functions of ISIs:

a. To assist existing and prospective entrepreneurs through technical and managerial counseling such as help in selecting the appropriate machinery and equipment, adoption of recognized standards of testing, quality performance etc.
b. Conducting Entrepreneurship Development Programs (EDPs) all over the country, on subjects like energy conservation, pollution control, technology up-gradation, quality improvement, material handling, management techniques etc apart from those involving facilities and assistance available from state and central government agencies, banks, financial institutions and national small industries corporation particularly for unemployed youth, ex-service personnel etc.
c. To advise the central and state governments on policy matters relating to small industry development.
d. To assist in testing of raw materials and products of small scale industries (SSIs), their inspection and quality control.
e. To provide market information to the SSIs.
f. To recommend SSIs for financial assistance from financial institutions.
g. To enlist entrepreneurs for participation in government stores purchase programme.
h. Conduct economic and technical surveys and prepare techno-economic feasible reports for selected areas and industries.
i. Identify the potential for ancillary development through sub-contract exchanges.
j. Organize seminars, workshops and industries clinics for the benefit of entrepreneurs.

7. **Khadi Village Industries Commission (KVIC) Schemes:** KVIC runs a number of schemes to support start-up/ new entrepreneurs. A brief of such schemes is enumerated below:

a. **Prime Minister's Employment Generation Program** (PMEG) is credit linked subsidy program, introduced after the merger of two already existing schemes i.e. Prime Minister's Rozgar Yojna (PMRY) and Rural Employment Generation Program (REGP). The major objectives of the scheme are discussed below.

- To generate employment opportunities in rural as well as urban areas of the country through setting up of new self employment ventures/ projects/ micro enterprises.
- To bring together widely dispersed traditional artisans/ rural and urban unemployed youth and give them self employment opportunities to the extent possible, at their place.
- To provide continuous and sustainable employment to a large segment of traditional and prospective artisans and rural and urban unemployed

youth in the country, so as to help arrest migration of rural youth to urban areas

- To increase the wage earning capacity of artisans and contribute to increase in growth rate of rural and urban employment.

The expected beneficiary contribution is 10% of project cost and subsidy rate is 15% for urban based ventures while 25% for rural based. Maximum project cost admissible under manufacturing sector is Rs. 25 lakhs while Rs. 10 lakhs under business/ service sector. The balance amount of the project cost will be provided by banks as term loan.

b. **Rural Industries Service Centre (RISC):** The aim of this program is to provide backward and forward linkages to khadi and village industry activities in a cluster and to provide services like raw material support, skill up-gradation, training, quality control, testing facilities, marketing promotion, design and product development in order to strengthen the rural clusters. It will be a common facility unit providing at least one of the following as discussed below:

- Testing facilities by establishing laboratory to ensure quality of products.
- Improved machinery/equipment to be utilized as common utility facilities by the nearby units /artisans to enhance production capacity or value addition of the product.
- Attractive and appropriate packaging facilities and machineries to the local units / artisans for better marketing of their products.

In addition to the above, RISC can also cater to the following services:

- Provide training facilities to upgrade artisan's skill in order to increase their earnings.
- Provide new design or new product, diversified product in consultation with experts /agencies for value addition of rural manufacturing units.
- Provide raw material support which mainly depends on seasonal procurement.
- Prepare product catalogue.

The industries covered under this scheme are khadi and poly vastra post weaving value addition facilities, herbal products (cosmetics and medicines), edible oil, detergents and soaps, hand-made paper, food processing, biofertilizers/ bio-pesticides, pottery, wood work, leather and others not in the negative list. Own contribution in this scheme is 25% (10% in North- eastern states) and KVIC's share is 75% (90% in case of NE states), subject to a limit of Rs. Five lakhs total project cost.

1. **National Small Industries Corporation (NSIC) Schemes:** A large number of schemes are also available with NSIC to start new enterprises. A glimpse of these schemes is presented below.

a. **Bank credit facilitation scheme**: To meet the credit requirements of MSME units NSIC has entered into a memorandum of understanding (MoU) with various nationalized and private sector banks. Through syndication with these banks, NSIC arranges for credit support (fund or non fund based limits) from banks without any cost to MSMEs.

   Furthermore the MSMEs can upgrade their competence in terms of business and technologies by getting rated through independent, renowned and professional rating agencies empanelled with NSIC. The empanelled rating agencies are CRISIL, ONICRA, SMERA-D&B, CARE, India Ratings, BRICKWORK and ICRA. The MSE which gets rated under **Performance and Credit Rating Scheme (NSIC)** not only has a liberty to get rated from any one of rating agencies of its preference but it also invariably increases their creditability in business and help them in getting timely credit from banks at liberal rates of interest. All documentations pertaining to completion and submission of a credit proposal to banks shall be undertaken by NSIC thereby saving cost and time to MSME.

b. **Performance and credit rating scheme**: Need of a performance and credit rating mechanism for SSIs (now micro and small enterprises) was highlighted in union budget of 2004-2005 (Table 11.1). A scheme for SSIs (now micro and small enterprises) has been formulated in consultation with Indian Banks' Association (IBA) and rating agencies. NSIC has been appointed the nodal agency for implementation of this scheme through empanelled agencies. Salient features of this scheme include:

   - A combination of credit and performance factors including operations, finance, business and management risk.
   - Uniform rating scale for all empanelled rating agencies.
   - SSIs (MSEs) have the liberty to choose among the empanelled rating agencies.
   - Turn-over based fee structure.
   - Partial reimbursement of rating fee through NSIC as follows:

**Table 11.1:** Performance and credit rating scheme of NSIC

| Turn Over of SSI /MSE | Reimbursement of Fee through NSIC |
|---|---|
| Upto Rs 50 Lacs | 75% of the fee or Rs 25000/- (Whichever is less) |
| Above Rs 50 to 200 lacs | 75% of the fee or Rs 30000/- (Whichever is less) |
| More than Rs 200 lacs | 75% of the fee or Rs 40000/- (Whichever is less) |

c. **Raw material assistance scheme**: This aims at helping MSEs by way of financing the purchase of raw material (both indigenous & imported). This gives an opportunity to MSEs to focus better on manufacturing quality products.

The scheme provides for financial assistance for procurement of raw material upto 90 days. MSEs are also helped to avail economics of purchases like bulk purchase; cash discount etc. NSIC takes care of all the procedures, documentation & issue letter of credit in case of imports.

2. **Small Industries Development Bank of India (SIDBI) Schemes:** A number of schemes are available from SIDBI to help start- up entrepreneurs. One of the schemes suitable for agri start-ups is **TIFAC –SIDBI Revolving Fund for Technology Innovation (SRIJAN scheme).** The scheme provides financial assistance to MSMEs towards development, up scaling, demonstration and commercialization of innovative technology based projects.

The assistance is given in the form of early stage debt funding on softer terms for development, demonstration and commercialization of new innovations in emerging technological areas, un-proven technologies, new products, process, etc. which have not been successfully commercialized so far.

Maximum assistance is generally not more than Rs. 100 lakh per project. Interest rate would be as approved by the project approval committee (PAC) (not be more than 5% per annum). All the proposals for assistance under the scheme shall undergo technical evaluation by Technology Information Forecasting and Assessment Council (TIFAC) and financial viability by SIDBI. The assistance is approved by a PAC consisting of officials of SIDBI and TIFAC.

The scheme was launched on 1$^{st}$ November, 2010 and aims at facilitating at development, demonstration and commercialization of technology innovation projects pertaining to new product or process development to encourage and promote development of capabilities in MSMEs to innovate and bring high risk innovations to the market for opening up opportunities for business linked with innovations. For further details, please see *https://www.sidbi.in/sites/default/files/products/TIFAC-SIDBIBrochure.pdf.*

3. **National Bank for Agriculture and Rural Development (NABARD) Rural Innovation Fund (RIF):** NABARD's RIF is designed to support innovative, risk friendly, unconventional experiments in farm, non-farm and micro-finance sectors that would have the potential to promote livelihood opportunities and employment in rural areas. The activities must have the rural in their focus and must be innovative, experimental and demonstrative in nature leading to replicability and commercial viability. The activities funded may involve development of new products, processes, prototypes, technology, patenting and extension support. Appropriate action research and studies contributing to better understanding of rural development issues, policy and process implementation may be undertaken.

All innovations and related activities in the farm, rural non-farm and micro-finance sectors can have access to the RIF. Assistance from RIF will be available for all activities which are in keeping with the guiding principles of RIF and specifically those which provide technology and skill up gradation, inputs supply and market support leading to promotion of viable enterprises, sustainable employment, infrastructure development, improved flow and access of credit to rural entrepreneurs. Innovations improving the efficiency of credit delivery and other support services to the rural resource poor and patenting innovations leading to commercialization of the idea through licensing or otherwise are also eligible for funding under this scheme.

The following areas/sectors will be considered for support from RIF- dry land / rain fed farming, innovative rainwater harvesting for rural dwellings, rural energy from biomass/ agricultural wastes, techniques for increasing the value of crop residues and non-crop bio mass, community regulation of distribution and use of water and energy, storage devices for agricultural and rural products, innovative methods of managing common property resources, materials and designs for rural roads and rural sanitation and waste disposal.

Apart from this, activities trying to bring in more innovative approaches/ ideas with greater focus in the following areas will also be supported: rural entrepreneurship/ skill development, marketing strategy and networking, rural housing and habitat development- low cost and natural calamity resilient housing structures for providing safe habitat to rural population, rural service sector, financial inclusion, micro-finance to micro-entrepreneurship, financial and technological model for health care and hygiene, evolving package of products on rural tourism and culture, rural sanitation and rural *haats* in collaboration with panchayat raj institutions. For more details, please see *https://www.nabard.org/english/rif_faqs.aspx.*

**11. SFAC Venture Capital Assistance Scheme:** The scheme of small farmers agri-business consortium (SFAC) aims:

a. To facilitate setting up of agri-business ventures in close association with all banks/financial institutions notified by Reserve Bank of India (RBI) where the ownership of the state/central government is more than 50% such as nationalized banks like SBI, IDBI, SIDBI, NABARD, National Cooperative Development Corporation (NCDC), North-eastern Development Finance Corporation Limited (NEDFi), Exim Bank, Regional Rural Banks (RRBs) and State Financial Corporations.
b. To catalyze private investment in setting up of agri-business projects and thereby providing assured market to producers for increasing rural income and employment.
c. To strengthen backward linkages of agri-business projects with producers.
d. To assist farmers, producer groups and agriculture graduates to enhance their participation in value chain through project development facility (PDF).
e. To arrange training and visits etc of agri-preneurs in setting up identified agri-business projects.
f. To augment and strengthen existing set up of state and central SFAC.

The SFAC venture capital scheme will assist agri-preneurs to make investments in setting up agri-business projects through financial participation and provide financial support for preparation of bankable detailed project reports (DPRs) through PDF. The SFAC will provide venture capital to qualifying projects on the recommendations of the bank/financial institution financing the project. This venture capital will be repayable back to SFAC after the payment of term loan of lending bank/ financial institution as per original repayment schedule or earlier. SFAC would provide venture capital to agri-business projects by way of soft loan to supplement the financial gap worked out by the sanctioning authority of term loan under means of finance with respect to cost of project subject to certain conditions. For details please see *http://sfacindia.com/PDFs/Revised-VCA_Operational-Guidelines.pdf.*

Further, in the union budget 2015-2016 announced on 28th Feb. 2015, setting up of Atal Innovation Mission (AIM) is envisaged in National Institution for Transformation of India (NITI), a replacement of planning commission, to provide innovation promotion platform involving academicians and drawing upon national and international experiences to foster a culture of innovation, research and development with an outlay of Rs. 150 crore.

Additionally, self employment and talent utilization (SETU) is being established as techno financial incubation and facilitation program to support all aspects of start-up business with an envisaged outlay of Rs. 1000 crore in NITI. In the union budget 2016-2017, start- up entrepreneurship has been given a unique focus. Most important one being three year tax holiday for start-ups.

**National Research and Development Corporation (NRDC)**

Techno-commercial support for promising innovations/inventions is available from NRDC (a Government of India enterprise) in the form of improvements in the prototype to make it commercially acceptable, carrying out process trials/ field trails/testing analysis, preparation of a pre-feasibility report, preparation of video or animation film about working of the invention, preparation of comprehensive know how document, technology demonstration (in deserving cases only) for a period of one year. The application can be made against call for proposals. The notification comes usually in the month of Jan-Mar every year. For more details please see *http://www.nrdcindia.com.*

**Techno-preneur Promotion Program (TePP) Project Fund (TFP) of Department of Scientific and Industrial Research (DSIR)**

Unattached independent innovators like micro and small entrepreneurs working on new designs and also the start- up firms/incubated firms with a turnover of less than Rs. 45 lakhs are eligible to apply for funds under this scheme. Proposals to convert an original idea innovation/know how into working prototype/process and so also the proposals to demonstrate novel delivery of science and technology innovations to rural India will be covered under the ambit of this scheme. Maximum support under this category is Rs. 15 lakhs subject to 90% of the approved project cost. The proposals can be submitted online (*http://oscar.iitb.ac.in/TEPP*) or submitted off line to the nearest TePP Outreach Centre (TUC). The innovators can meet coordinators of TUCs and take their advice before submitting the application. Addresses of TUCs are given on *www.dsir.gov.in.*

**National Horticulture Board**

National Horticulture Board established in 1984 offers two major schemes to horticulture entrepreneurs, one is on production and processing of horticulture crops and the other is for cold storages. Details of these two schemes are given below:

1. **Scheme for development of commercial horticulture through production and post-harvest management of horticulture crops.**
   a. **Production related:** It offers back ended capital subsidy of 20% of the total project cost upto Rs. 25 lakhs per project (Rs. 30 lakhs in North Eastern, hilly and scheduled areas). However for capital intensive and high value crops under protected cultivation and open air cultivation of date palm, olive and saffron, subsidy will be @ 25% of project cost with a ceiling of Rs. 50 lakhs (33% of project cost with a ceiling of Rs. 60 lakhs for scheduled and hilly areas). The eligible projects include:
      - High quality commercial horticulture crops
      - Indigenous crops/produce, herbs, spices
      - Aromatic and medicinal plants
      - Seed and nursery
      - Biotechnology, microbiology, biochemistry, biodiversity and tissue culture
      - Protected cultivation
      - Biopesticides
      - Organic fertilizers, organic food, bio-dynamic farming, vermi-compost
      - Establishment of horti health clinics/labs
      - Hydroponics
      - Aeroponics
      - Bee keeping and its products
      - Mushroom and its products
      - Nuts and their products.

   b. **Processing related:** Credit linked back-ended subsidy @ 40% of the total project cost limited to Rs 50 lakh per project in general area and @ 55% of project cost limited to Rs 60.00 lakh in hilly and scheduled areas. Pattern of assistance for plastic crates will be 50% of the total cost. The eligible projects include:
      - Washing, drying, sorting, grading, waxing, packing palletizing, freezing units etc.
      - Pre cooling units/cool chambers
      - Refer van/containers
      - Specialized transport vehicle

- Retail outlets
- Auction platform
- Ripening/curing chamber
- Market yards/rope ways
- Irradiation/vapor heat treatment (VHT) unit
- Primary processing of products (fermentation, extraction, distillation, juice vending, pulping, dressing, cutting, chopping, dehydration etc)
- Natural color and dye extraction
- Essential oils, perfumery, cosmetics out of horticulture products
- Products out of horticulture waste
- Horticulture ancillary industry for promoting indigenous manufacturing of horticulture related farm tools and machineries, equipments, plastic containers, packaging etc
- Adoption of quality assurance system (HACCP, TQM, ISO, Global Gap etc)
- Plastic crates and bins, cartons, aseptic packaging and nets.

2. **Capital investment subsidy scheme for construction/expansion/ modernization of cold storages/storages of horticulture produce:** The scheme includes credit linked project related to cold storages including controlled atmospheric (ca) storage, modified atmospheric (ma) storage, pre cooling units, other storages for onion and their modernization.

The assistance will be as credit linked back-ended subsidy @ 40% of the capital cost of project in general areas and 55% in case of hilly & scheduled areas for a maximum storage capacity of 5000 MT per project. The general conditions for cold storage projects are discussed as below.

**a.** Multi-chamber cold storages with mezzanine floors of RCC and or wooden structure for products not requiring pre-cooling, temperature range 0°C to 16°C or above with fin-coil cooling system, with control system for humidity (RH 80% to 95% in general and 65%-70% for onion & garlic or appropriate level for seed storage etc) and appropriate handling of stored commodity with safety system against accidents of fire, leakage of refrigerant etc. Minimum of two chambers, standard insulation material, with civil structure, insulations and cooling system as per prescribed standards - @ Rs 6,000 per MT

**b.** Multi-chamber and multi-product cold storage for whole range of horticulture produce but without pre-cooling system (temperature range from -2°C or lower to +16°C plus (with temperature, humidity and $CO_2$ control (RH= 80% to 95% for F &V in general and 65%-70% for onion & garlic or appropriate level for seed storage etc), fin-coil cooling system and energy saving devices for peak and lean load period, heat exchanger with $CO_2$ control system, appropriate technology for control of microorganism / dust mite and mechanical handling of stored commodity such as racks, pallets & forklift / stacker (as per site conditions); bags/ bins and crates/ CFB boxes and also safety devices against accident.

**c.** With civil structure/pre-fabricated engineering structure, insulation, cooling system etc as per prescribed standards excluding cold storage with mezzanine floors @ Rs 7000/- per MT.

**d.** With civil structure/pre-fabricated engineering structure, insulation, cooling system etc. as per prescribed standards (excluding cold storage with mezzanine floors) and having on-farm or off-farm pack house facility of washing / de-sapping (as per requirement), drying, sorting, grading, waxing, packing etc and pre-cooling @ Rs 8000/- per MT. Modernization of cold storages:

- Up gradation of thermal insulation.
- Up gradation of cooling system, refrigeration, air flow, electric installation, handling devices, safety devices etc.

**Funding and other support from Villgro Innovation Foundation**

Villgro is India's oldest social enterprise incubator based out of Chennai, which has supported 103 enterprises, secured Rs. 873 million in follow on funding and touched over 6.8 million lives. Villgro supports innovations impacting bottom of pyramid (BoP) markets including those in agriculture. For example Villgro has supported fly bird innovations which is an affordable irrigation controller that helps farmers increase yield and quality. Another start- up supported by Villgro is Promethean which is possibly the first rapid milk chilling machine deployed in villages of India with leading dairy companies. Others include Kamal Kisan (range of agricultural implements for small and marginal farmers). Under the Mango Tree (unique range of honey for the consumers and bee keeping solutions that improve farm productivity), Skymet (private weather forecasting and dissemination information to farmers via mobile phones), Masuta (first tsar yarn making machine that can increase yield several times), IWAC (areca nut de-husking machine that can increase output per unit time by six fold) and Aroghyam (organic value chain that benefits both consumers and farmers). Villgro support is available in the form of funds, mentoring, talent and networks. To submit an innovative idea, please click *http://www.villgro.org/incubation-apply.*

**National Innovation Foundation (NIF)**

NIF has initiated a micro venture innovation fund (MVIF) of Rs 4 crore for ten years with the support of SIDBI in October 2003, which got operationalized in January 2004. Under the scheme, support from MVIF is made available to innovators and entrepreneurs who are associated with NIF for technology commercialization.

The uniqueness of MVIF of NIF is the first and only of its kind micro venture risk fund in the world, which extends financial support to grassroots innovators under a single signature on a simple agreement of understanding without any collateral or a guarantor. Unlike micro finance or conventional venture funds, MVIF invests in risk areas with high failure probability, investing in those technologies and products for which either market does not exist or may be very limited. Apart from financial returns, one of the key criteria for selection of a technology for MVIF support is social return or social value created for the benefit of society at large.

NIF extends the MVIF support to the innovators directly as well as through their regional incubators i.e. GIAN and GIAN Cells. The team at business development (BD) department of NIF conducts initial market assessment and makes the investment plan and the proposal, which is presented to the fund management committee (FMC). The FMC evaluates and gives recommendations on the proposals submitted by BD team. Under MVIF, on case to case basis, funds are offered with reasonable moratorium period with monthly and also seasonal repayment with consideration of about 12.5% interest. The approval procedure is unique, fast and hassle free for the innovator as we are not demanding innovators to submit anything except simple request about their requirement. The NIF team does all documentation and provides mentoring and all kind of hand holding support to the innovator. The FMC also advises the BD team about project monitoring, commercialization strategy and suggests mid-term corrections if any. A local mentoring team also facilitates proper utilization of funds and guides the innovator from time to time. Out of total four crore fund, till 2014-2015, total 192 projects have been supported with total disbursed amount is Rs. 3,40,97,637 and total repayment amount being Rs. 2,16,37,496.

**Global Social Venture Competition (GSVC)**

It provides aspiring entrepreneurs with mentoring, exposure, and $50,000 in prizes to transform their ideas into businesses that will have positive real world impact. Founded in 1999 by MBA students at University of California Berkeley's Haas School of Business, the GSVC culminates each year with the global finals and Conference at Berkeley in April, gathering teams from around the world and bay area professionals for a day of learning and networking. First regional finals are held on a generally continent basis. GSVC has evolved into a global network supported by an international community of volunteer judges, mentors and student organizers and a partnership of premier business schools in the Americas, Europe, Asia, and Africa. The announcement for preliminary proposal submission generally comes in October-November each year. An important condition to submit the b-plan/proposal is that one of the team members should be an MBA student. For details, please see *http://gsvc.org*.

**Bank Schemes (Indicative only)**

Banks in India cater to agri-preneurs through offerings like agriculture loans or SME loans. The schemes vary from bank to bank in terms of the quantum of offering, security requirements, interest rates, activities financed etc. An indicative brief of the bank schemes for agri start ups/ agri-preneurs is give below:

1. **Bank of Baroda's** technology up-gradation fund scheme for textile and jute industries: BoB grants loans under technology up-gradation fund scheme launched by Government of India as per guidelines received from time to time from Ministry of Textiles. BoB is a nodal agency for determining eligibility and releasing of subsidy for the cases financed by the bank under

the scheme. The objective is to provide encouragement to textile industrial units for taking up technology up-gradation and modernize their production facilities. The scheme envisages 5% interest reimbursement (4% for spinning industry) of the normal interest charged by the bank on the loans availed by the units from the bank for undertaking technology up-gradation.

2. **Canara bank** offers upto 85% loan for starting units in tissue culture, floriculture, mushroom cultivation, aquaculture, export oriented agriculture, seed production and processing, lift irrigation projects, piggery, rabbit rearing, bee keeping, scientific sheep and goat rearing, bio gas plants, purchasing fleet of tractors and well boring equipment for custom renting, farm forestry, bio-diesel plants, tree borne oil seeds etc. upto 85% of project cost. Rest of the amount will be borrower's contribution as margin money. Wherever subsidy is available, it can be treated as margin money.
3. **Axis bank** provides loans for contract farming. Both pre and post harvest finance is extended to the producer. In turn farmer mitigates marketing risk by tie up with a reliable partner (corporate). He also gets an incentive for growing premium quality crop variety. For details, please see *http://www.axisbank.com/agri-rural/agri-business-loans/contract-farming/contract-farming.aspx or visit nearest Axis bank branch.*

**Training and Capacity Building**

Training and capacity building is yet another indispensible requirement of start-ups. It may pertain to procedures to start a new enterprise to certifications and licenses to compliances of the regulator. There are a number of organizations in India which provide entrepreneurship training to start-ups. A brief description of such organizations and their offerings is given below:

1. Entrepreneurship Development Institute (EDI) of India, Ahmedabad offers a two years Post Graduate (PG) diploma in business entrepreneurship (with a seminar course on agri-business) and another two years PG Diploma in Development Studies. It also offers an open learning program –Diploma in Entrepreneurship and Business Management through distance mode. Apart from this it offers a number of short term training programs for needy start-up entrepreneurs. The details of such short term training programs can be had from the website of the institute *http://www.ediindia.org*.
2. **National Institute of Entrepreneurship and Small Business Development (NIESBUD),** Noida (with branch office at Dehradun) is engaged in training, consultancy and research in order to promote entrepreneurship and offers short term training programs in agricultural water lifting equipments repairing and maintenance, bakery products, biotechnology, dairying, essential oil and perfumery products (125 hours), food processing (150 hours), mushroom cultivation (125 hours), landscape design (125 hours) and management development programs (MDPs) on carbon trading, export procedures, documentation and management.

3. **Indian Institute of Entrepreneurship (IIE),** Guwahati established in 1993 is an organization of the Ministry of MSME and undertakes training, research, consultancy activities in small and micro enterprises mainly in the North Eastern region. It conducts a two days MDP on marketing of rural products for NGOs. It also conducts a capacity building program on post harvest management and entrepreneurship. Please see *http://iie.nic.in*.
4. **National Institute of Micro, Small and Medium Enterprises (NI-MSME),** Hyderabad established in 1960, works for promotion, development and modernization of the SME sector. It conducts a training program of around eight week duration on planning and promotion of agro and food enterprises (PAFE). The program mainly focuses getting the participants acquainted with industrial possibilities from the available agro and allied resources, sharpening professional skills to formulate and appraise agro and food projects, graining exposure to agro and food products and international quality standards for trade and exports and gaining from Indian experience to evolve suitable action plans for stimulating growth of agro and food enterprises in back home situation. Please see *www.nimsme.org*.
5. **National Institute of Food Technology Entrepreneurship and Management (NIFTEM):** NIFTEM plays a critical role in addressing the issue of talent gaps in the food processing industry by way of continuing education and training programmes. NIFTEM as an apex institute in food science and technology is a harmonizer in bringing about co-ordination between the industry, academia and government to create the required ecosystem for skill development to fill the talent gaps through training. NIFTEM's continuing education focuses on the skills deficit in food processing industries by devising and delivering innovative training solutions. The institute conducts following training programmes under continuous education.

   a. **Executive development programmes** (EDP) are designed for food processing professionals for updating knowledge in various sectors of food processing.

   b. **Entrepreneurship development programmes** to foster the entrepreneurial spirit and equip the enrolled budding entrepreneurs in getting a foot hold in the industry.

   c. **Skill development programmes** for updation of knowledge and skill enhancement of professionals at different levels of industry across various sectors particularly for:

      - Production managers and engineers
      - Quality control and R&D specialists
      - Supply chain & logistics professionals
      - Regulatory & legal experts
      - Shop floor technicians, refrigeration mechanics, supervisors & others.

6. **National Institute of Agricultural Extension Management (MANAGE), Hyderabad** runs a National Bank for Agriculture and Rural Development (NABARD) funded agri-clinics and agri-business centres (AC& ABCs) scheme of Government of India wherein it trains unemployed agri and allied/life science graduates in all aspects of agri entrepreneurship for two months. Additionally, after this course, the beneficiaries are also eligible for bank loans to set up their enterprise.
7. **Industrial Technical Consultancy Organizations:** Most of the states have industrial technical consultancy organizations (Table 11.2). For example in Gujarat, there is Gujarat Industrial Technical Consultancy Organization (GITCO) with head office at Ahmedabad which basically offers services like project opportunity identification, preparation of techno economic feasibility reports, project appraisal, technology evaluation studies, patent assistance and organizing seminars and conferences. Please visit *http://www.gitco.co.in.* Similarly in Punjab, there is North India Technical Consultancy Organization Limited, popularly known as NITCON (*www.nitcon.org)* which offers an array of services including techno economic feasibility reports preparation for investment proposals envisaging green field projects as well as those of expansion, modernization and diversification etc., market survey and assessments, entrepreneurship and skill development training programs.

**Table 11.2:** Services offered by some of the state technical consultancy organizations (TCOs)

| S. No. | Name of TCO | Major services offered* | Website/link |
|---|---|---|---|
| 1 | Andhra Pradesh Technical Consultancy Organization (APITCO) | Food and Agri-business management & human resource consulting | www.apitco.org |
| 2 | Himachal Consultancy Organization Limited (HIMCON) | Valuation, techno economic feasibility studies, surveys, project appraisals and skill development training programs | http://himcon.org |
| 3 | Kerala Industrial and Technical Consultancy Organization Limited (KITCO) | Asset valuation, techno economic feasibility reports (TEFRS), market surveys, detailed project reports (DPRS), debt restructuring, skill development and certification, training, entrepreneurship development etc. | www.kitco.in |

| S. No. | Name of TCO | Major services offered* | Website/link |
|---|---|---|---|
| 4 | Technical Consultancy Services Organization of Karnataka (TECSOK) | Preparation of DPRs, project appraisals, enterprise restructuring, evaluation studies, valuation of assets, environment related studies etc. | http://tecsok.com/ |
| 5 | Madhya Pradesh Consultancy Organization (MPCON)-also caters to Chhattisgarh | Surveys, trainings, project reports, management consultancy, development of women entrepreneurship and allied activities, symposia and seminars. | mpcon.tripod.com/ |
| 6 | Maharashtra Industrial and Technical Consultancy Organization Limited (MITCON) | Agri infrastructure and food consulting, biotech and pharmaceutical, textile and clusters infrastructure consulting, quality assurance. | www.mitconindia.com/ |
| 7 | Uttar Pradesh Industrial Consultants (UPICO) | Biotech and pharma entrepreneurship training and consultancy | http://www.upico.org/ |
| 8 | Webcon Consulting India Limited (formerly West Bengal Consultancy Organization Limited (WEBCON) | Techno- economic studies, market survey/research, organization restructuring, industrial potential survey, micro enterprise development, cluster development program, entrepreneur nurturing | |
| 9 | ITCOT# Consultancy and Services Limited (Tamilnadu) | Feasibility reports, project appraisal, DPRs, business plan for venture/angel finance, market research/ consumer research, entry strategies, evaluation studies, sectoral reports | http://www.itcot.com |

*Services related to agri entrepreneurship only are shown
#Industrial Technical Consultancy Organization of Tamil Nadu (former name)

## Scientific Expertise

Depending upon the idea/concept, the scientific expertise requirement of the start-up may vary from low to very high. Today, it is available in State Agricultural Universities (SAUs). There are subject/crop specific research institutes of Indian

Council of Agricultural Research (ICAR). Please see *www.icar.org.in.* Many Council of Scientific and Industrial Research (CSIR) laboratories are working in agriculture and provide scientific expertise to needy start-ups with their own mechanisms like business incubation. Many of the ICAR research institutes like National Dairy Research Institute (NDRI) and National Academy of Agricultural Research Management (NAARM) have a structured mechanism of business incubation to support agri start-ups. These business incubators are initially funded by National Science and Technology Entrepreneurship Development Board (NSTEDB), Department of Science and Technology (DST), Government of India. The concept of business incubation will be discussed in detail in later section in this chapter.

## Product Testing and Trial

There are many laboratories of national importance mostly in the public sector which carry out the testing and trail of agri and food products at nominal rates. These facilities can be availed by the start-up agri entrepreneur wishing to get his/her product tested. A few examples of such labs are Centre for Analysis Learning in Livestock Food (CALPH) at National Dairy Development Board (NDDB), Anand. There are few private laboratories also doing such analysis but their charges may be higher. Examples of such private laboratories are Xcelris Labs (an Abellon Company), Ahmedabad, SGS and Reliable Analytical Laboratories. Some of the SAUs and ICAR Research Institutes are also having world class product testing laboratories. For example Pesticide Residue Testing Laboratory (ICAR Unit) at Anand Agricultural University is NABL Accredited and carries out analysis of agrochemical residues in fruits, vegetables and other such products.

## Mentorship and Advisory

The major difference between a mentor and a consultant is that a consultant provides solution to the problem but a mentor takes us out of the problem. Today, a large number of mentors are available to agri start-ups (Table 11.3). Many of them are working professionals, other are practicing entrepreneurs themselves, yet others are retired scientists/managers etc. The charges of mentors are based on time, number of visits, solutions provided. They seldom ask for equity stake in the company. But this could be considered an option, if mentoring becomes unaffordable for the start up/innovator.

**Table 11.3:** Indicative list of mentors, their contacts specialization areas

| S. No. | Name of mentor and Affiliation (if any) | Area of Specialization | Contacts |
|---|---|---|---|
| 1 | Dr. Sridhar Bedarkar, Brilliant Bio-pharma, Hyderabad | Veterinary, dairy & food business, international trade | Ph: 9848041875,9921197291<br>Email: nbshridhar@gmail.com, drshridhar12@rediffmail.com |
| 2 | T. Sivakumar, Rasim Exports, Chennai | Textiles, garments, merchandising, materials management | Ph: 91 9884171753<br>Email: t.sivakumar80@gmail.com |

| S. No. | Name of mentor and Affiliation (if any) | Area of Specialization | Contacts |
|---|---|---|---|
| 3 | J. Siva Kumar, Land Catch Farms, Federal Republic of Nigeria | Fish breeding and rearing, aquaculture probiotics | Ph: 09444249081, 9176813747<br>Email: sivajeyam@yahoo.com |
| 4 | K. Devdas Nair, Thrichur, Kerala | Marketing, brand awareness, leadership development | Ph: 91-9747953165<br>Email: devadasnair@rediffmail.com |
| 5 | Dr. Suresh Devatkal, Senior Scientist, PAU, Ludhiana | Livestock production and management | Ph: 91-9417460662<br>Email: sureshlpt@yahoo.com; skdevatkal@gmail.com |
| 6 | K. Vinod Shenai, Nilgiris, Tamilnadu | Tea, coffee, cardamom cultivation and processing, eco-tourism | Ph: 91 9444993173<br>Email: vinod.shenai@gmail.com |
| 7 | B. Issac Arul Jeyaprasath, Coimbatore | Quality control and TQM | Ph: 91-9790666697<br>Email: sscbacas@gmail.com |
| 8 | Anil Chopra, MD, Competent Agri Solutions, New Delhi | Dairy, fruit and vegetable processing | Ph: 91-9811814559, 9868164559<br>Email: anil.chopra@agrisolutions.in |
| 9 | Dr. Tamma Gangadhar Rao, CAIIB, Mumbai | Animal health/ animal nutrition/management of dairy plant/ poultry farming and hatchery; commercial poultry farming, dairy farming, sheep rearing and small scale industries sales and procurement of milk and milk products/ processing of milk/ management of milk co-operative societies | Ph: 0884- 2350580<br>Mobile: +91-9000777975<br>Email: grtamma@yahoo.com |
| 10 | Kanan Narayanswami, Founder Start- up to scale up, Chennai | Funding, investing and incubating, mergers, acquisitions, internal process Monitoring, scaling up | Ph: 91-9444402605<br>Email:kannannarayan@gmail.com |

## Pilot Production

It would be a costly affair for a start-up entrepreneur to set up the entire production plant only to realize later that product doesn't sell or the set up is not viable. Considering this, it would be a boon for the start-up, if he/she gets a pilot production facility (on rental) to manufacture few batches/samples of the product for test marketing. But this has two major limitations. If the product is entirely new to the world, there won't be any existing production facilities available. Secondly, private entrepreneurs would not be ready to offer their production facility for sample production on rental. Under such circumstances, the only option left with the entrepreneur is to approach contract/custom research labs. However, they are very few and expensive. Some possibility of getting pilot production done for agri based products exists in the pilot plants of state agricultural universities running in various schemes like experiential learning.

### Design Validation

The only institute of national importance working on design in India is National Institute of Design (NID) at Ahmedabad. A product definitely needs a design before being introduced in the market. Design holds high importance in agricultural engineering products, tools, equipments and other such gadgets. They also have a design business incubator. For details, please visit *http://www.nid.edu.*

### Test Marketing

Test marketing a product is another challenge for a start-up. A market study may be required even before test marketing of the product. A start-up entrepreneur may not have the necessary resources to carry out market study. If he outsources the market study to an agency/consultant, again it may be digging too deep in his pockets.

### Procedure to Start a New Business (Legal Aspects)

In India, incorporation of a company is governed by the Companies Act 1956. It is the most important piece of legislation that empowers the central government to regulate the formation, financing, functioning and winding up of companies. It applies to whole of India and to all types of companies, whether registered under this act or an earlier act. But it does not apply to universities, co-operative societies, un-incorporated trading, scientific and other societies.

For registration and incorporation of a company, an application has to be filed with registrar of companies. Application for registration of a company accompanied by the selected names, memorandum of association and articles of association and other necessary documents is to be filed with the registrar of companies of the State in which the company is proposed to be incorporated.

Under the companies act, an entrepreneur can form two types of companies, namely a private company or a public company. A **private company** is one, the articles whereof contains the following restrictions as discussed below:

1. Restricts the minimum paid up share capital to such an amount as may be prescribed but which shall not be less than rupees one lakh;
2. Restricts the rights of members to transfer its shares, if any;
3. Limits the number of its members to fifty excluding the past or present employees of the company who are members of the company;
4. Prohibits any invitation to the public to subscribe for any shares or debentures of the company;

5. Does not invite or accept any deposits from persons other than its members, directors or their relatives. Also, the minimum number of members in a private company is two and such a company must have the words 'Pvt Ltd' as the last part of its name.

A **public company**, as defined in the companies act, has the following features as discussed below:

1. Its shares are freely transferable;
2. There is no ceiling on its membership;
3. It can invite general public to subscribe to its shares;
4. It has a minimum paid up capital of Rs.5 lakhs or such higher paid up capital as may be prescribed;

It is a private company which is a subsidiary of a public company. Also, the minimum number of members in a public company is seven and such a company must have the word Ltd as last part of its name.

A business enterprise can be owned and organized in several forms. Each form of organization has its own merits and demerits. The ultimate choice of the form of business depends upon the balancing of the advantages and disadvantages of the various forms of business. The right choice of the form of the business is very crucial because it determines the power, control, risk and responsibility of the entrepreneur as well as the division of profits and losses. Being a long term commitment, the choice of the form of business should be made after considerable thought and deliberation. The choice of the form of business is governed by several interrelated and interdependent factors are discussed below.

The nature of business is the most important factor. Businesses providing direct services like tailors, restaurants and professional services like doctors, lawyers are generally organized as proprietary concerns. While, businesses requiring pooling of skills and funds like accounting firms are better organized as partnerships. Manufacturing organizations of large size are more commonly set up as private and public companies.

Scale of operations i.e. volume of business (large, medium, small) and size of the market area (local, national, international) served are the key factors. Large scale enterprises catering to national and international markets can be organized more successfully as private or public companies. Small and medium scale firms are generally set up as partnerships and proprietorship. Similarly, where the area of operations is wide spread (national or international), company ownership is appropriate. But if the area of operations is confined to a particular locality, partnership or proprietorship will be a more suitable choice.

The degree of control desired by the owner(s). A person, who desires direct control of business, prefers proprietorship, because a company involves separation of ownership and management. Amount of capital required for the establishment and operation of a business. A partnership may be converted into a company when it grows beyond the capacity and resources of a few persons. The volume of risks and liabilities as well as the willingness of the owners to bear it is also an important consideration. Comparative tax liability.

## Regulatory Requirements

Once an entrepreneur has taken all the important decisions relating to starting a business, he/she has to take into account the basic regulatory requirements which are to be followed for setting up the organization. The most important regulation is the Companies Act, 1956, which regulates all the affairs of a company. It contains provisions relating to the formation of a company, powers and responsibilities of the directors and managers, rising of capital, holding company meetings, maintenance and audit of company accounts, powers of inspection and investigation of company affairs, reconstruction and amalgamation of a company and even winding up of a company. The Ministry of Corporate Affairs, (earlier known as Department of Corporate Affairs under Ministry of Finance), is primarily concerned with administration of this act as well as other allied acts and rules & regulations framed.

The next important regulation relates to environment. The environmental regulatory requirements envisage a wide legislative framework covering every aspect of environment protection like air, water, noise, forest conservation, wildlife protection, etc. Also, separate set of laws and rules for emission of hazardous wastes have been enacted. The Ministry of Environment and Forests (MoEF), is the nodal agency for regulating all such environmental aspects. It undertakes conservation and survey of flora, fauna, forests and wildlife; prevention & control of pollution; afforestation and regeneration of degraded areas. Every industry has to abide by all such guidelines and parameters for environmental protection because only this will ensure its sustainable progress and growth.

## Agri-business Incubators-One stop shop for Agri biz start-ups

In the earlier sections, we have seen that a start-up's life is not easy. Solutions to his problems are not only distant but also expansive. In many cases solutions do not even exist. This is where agri-business incubators come very handy to the start-up. Agri-business incubation is a new concept in India; however, the model is quite old in western world. Agri-business incubators principally work on the lines of general/ technology business incubators, with the difference being that they incubate only agricultural/agri-business start-ups (Table 11.4). The other difference is that business of agriculture itself is highly different from other business like manufacturing and services as agriculture is dependent on nature and predicting nature is becoming more and more difficult in the age of climate change. There may be mixed kind of business incubators with one of the portfolios being that of agri start-ups. A listing of such incubators dealing with agri start-ups is presented in next page.

**Table 11.4:** Business incubators with stakes in agricultural / agri-business sector

| S. No. | Name of business incubator | Year of set up | Main sub-sectors of work in Agri-business domain |
|---|---|---|---|
| 1 | Society for Innovation and Entrepreneurship in Dairying, National Dairy Research Institute, Karnal | 2008 | Dairying, fisheries, honey and bio based agri inputs |
| 2 | Composite Technology Park, Bangalore | 2008 | Composites based on coir, bamboo and jute |
| 3 | MITCON Biotechnology Incubation Centre, Pune | 2002 | Agricultural biotechnology |
| 4 | BIT-TBI, Sathyamangalam, Tamil Nadu | 2007 | Agricultural biotechnology |
| 5 | Centre for Innovation, Incubation and Entrepreneurship (CIIE), IIM Ahmedabad* | 2002 | Agri and allied sectors |
| 6 | Agri-business Incubator (ABI), International Crops Research Institute for Semi Arid Tropics (ICRISAT), Hyderabad | 2003 | Agriculture and Agri-business |
| 7 | Association for Innovation Development of Entrepreneurship in Agriculture (A-IDEA), National Academy of Agricultural Research Management (NAARM), Hyderabad | 2014 | Agriculture and Agri-business |
| 8 | Directorate of Agri-business, Tamil Nadu Agricultural University (TNAU), Coimbatore | 2007 | Agriculture and Agri-business |

*CIIE Initiatives started in 2009 *Source: Author*

Off late Ministry of MSME also plans to set up incubators to promote small enterprises including those in the agriculture and agri-business sector. Please see box 2.

**Role of an agri-business incubator**

The role of ABIs can be classified into five broad categories as follows:

**1. Promotion of agri innovations:** Gone are the days when farmers were mere beneficiaries of top-to-down dissemination of technologies. Days have come to recognize bottom-up innovations and of equal partnership of farmers in agricultural research and development. Farmers and rural people while working in agriculture keep on innovating new methods and procedures to solve their day to day problems. These innovations could range from very simple ones like staking of tomato plants to avoid touching of fruits with soil and increase aeration so as to reduce fruit rot many other diseases to development of altogether new hybrids/varieties of crops by simple selection or other complex methods. When a rural folk invents something new which may have a commercial value in the future, he may not necessarily have the resources in terms of money, scientific consultancy, machines and instruments to take his innovation to the market.

Here the role of an agri-business incubator comes very handy to the innovator. Most of the banks would not be ready to fund development of any innovation which has not seen the light of the day in terms of a big market. ABIs help arrange funding from various sources to take these innovations to the market. A number of possible funding sources for upstart-ups have been discussed earlier in this chapter.

**2. Promotion of agri-preneurship:** Progressive farmers or other individuals who want to start their own agri- business enterprises are supported by ABIs. ABIs also support start-up agri-entrepreneurs. Many corporate who diversify into new businesses are now a day's taking incubation support from ABIs. This support may come in the form of helping them taking benefits of government subsidies available to start new enterprises as those by NABARD, MSME, SFAC, preparation of business plan, scientific expertise available in the host institute, testing and trial facilities in its laboratories, arranging for mentors to the entrepreneurs according to their business areas. Mobilizing funding for them in terms of venture capital, seed funding etc by bringing the investor and entrepreneur face to face at relevant forums, preparation of techno-economic feasibility/viability reports etc.

**3. Agro-technology commercialization:** Since most of the ABIs are situated in educational or research institutes, they have a very good access to the technologies generated in these institutes. But all the technologies emerging from these institutions may not viable in terms of market potential, demand by the farmers/end users, returns and break even point. Hence, all the technologies need to be sorted out on various criteria. This is called evaluation of the technologies. All technologies are evaluated on three parameters. These are technical attributes (innovativeness, compatibility with the farming systems, complexity, process advantage, developmental maturity, benefits to end user, future scope for improvement, technical expertise available and technology readiness level), business attributes (market demand, business opportunity, revenue potential, time to reach market, competitive advantage, entry barriers, exit barriers, cost advantage, geographical market reach, regulatory acceptability, public perception) and social attributes (benefit to farmers, employment opportunities, benefits to society, impact on environment and life forms). This method of technology evaluation is called strategic assessment method (SAM). Another method of technology evaluation is called analytic hierarchy process (AHP). In this approach, all the selected factors are weighted equally for technology assessment. Relative scores are given for technologies and total scores obtained by the technologies are ranked and considered accordingly for commercialization. AHP can be used to relatively rank technology ideas with respect to various criteria, moving into or out of a depth of hierarchy as and when needed. The framework ensures that weights and decisions will remain consistent right through.

**4. Consultancy:** The scientific and technical consultancy is another offering from ABIs. Any agri based start-up wishing to have technical consultancy for his/her innovation may seek the support of an agri-business incubator. The incubator manager facilitates a meeting of the start-up and the concerned scientist in the parent institute. Upon mutual consent, an agreement may be signed between the start and the ABI highlighting the terms of reference of the consultancy service assignment including the charges and payment schedule.

In general, agri-business incubators have a facilitative role but this eases a lot of burden off the start up's shoulders. ABIs also maintain a mentor's directory to help the start-ups. These may not be the research scientists of the parent institute but freelancers, practicing entrepreneurs or retired technocrats. Often they charge a nominal charge for their services and rarely can also provide it on a pro bono basis. All agri-business incubators of a country generally have an association/ network. The power of network is that if a start- up approaches a particular agri business incubator and that particular incubator is not able to provide the required support, then he may directed to the other incubator (s) which can address the issues being faced the concerned start up.

To conclude, agri-business incubators are proving to be a one stop shop for innovators/start-ups in agri-business. The dire need of funding support to start-ups has also been addressed by ABIs though most ABIs do not directly invest in incubate companies but they can help establish close contacts of incubatees with angel investors and venture capitalists to an extent that start-up gets enough choices to present his B-plan to a number of investors. Other services like detailed project report, facility rental, support for test marketing, pilot production are also efficiently provided by agri-business incubators. Specialized and sector specific incubators like National Design Business Incubator at National Institute of Design (NID), Ahmedabad. Connecting with mentors and technical experts are other services successfully provided to start-ups by agri-business incubators. A great advantage of being incubated at Technology Business Incubator (TBI) or Science & Technology Entrepreneurship Park (STEP) recognized by National Science and Technology Entrepreneurship Development Board (NSEDB), Department of Science and Technology (DST) is that the incubate companies are exempted from service tax on the services which they render from TBI or STEP.

**Box 2**

***Government to set up 10 incubators to promote entrepreneurs***

By Express News Service – Bhubaneswar
Published: 06th July 2013 12:47 PM

The state government has firmed up plans to set up 10 small and medium enterprises (SME) incubators to nurture and promote innovative entrepreneurial ventures within the current year. The incubators would be funded by the Union MSME Ministry and hosted by technological universities and institutions. While Odisha's first and only business incubator at KIIT University, supported by Union Department of Science & Technology, is functional since 2009, the State MSME department has got on to the job of identifying other institutions for opening such facilities. The objective is to provide a proper environment with high-class infrastructure, laboratories, workshops, libraries and resource materials to enable the budding entrepreneurs to give shape to their ideas and innovations. The incubators would offer complete support right from fostering and developing the ideas into enterprises to ensuring successful commercialization of the ventures. The target is to identify and include at least 10 incubatees in the initiative in the first year. They would be provided with all the hand-holding support to make their ideas develop into self sustaining businesses, Secretary, MSME Department Panchanan Dash said. The department is also set to push barcoding of MSME products manufactured in the state in order to broadbase their acceptability and marketability. A whole lot of products are failing to find markets due to lack of barcodes, which have become the basic standard for manufactured output across the globe. The SME products are also not finding place in the malls and retail outlets that are flourishing across the town and cities owing to non-barcoding. The centre is providing subsidy to the tune of 75% of the cost of registration of barcode for a company. We will be assisting enterprises to register for barcodes and gain the competitive edge in the market. Around 15 enterprises would be assisted for barcoding during this year, Dash added. The MSME department has come out with an ambitious action plan for 2013-14 that sets a target of producing 10,000 entrepreneurs from the unemployed youths in next one year by holding entrepreneurship development programmes in nook and corner of the State. The youths will be provided skill training and hand-holding support from project report preparation, registration to credit linkages and establishment of ventures. It will also take up two industrial clusters for implementing the lean manufacturing competitiveness scheme for MSMEs. The scheme focuses on implementing techniques to identify and eliminate waste in the manufacturing processes, thereby effecting cost-optimization and manufacturing competitiveness.

## Further Readings

1. http://www.publishyourarticles.net/eng/articles/what-are-the-main-functions-of-small-industries-service-institute-sisis.html accessed on 6th March 2015 at 10:46 PM IST.
2. https://www.sidbi.in/sites/default/files/products/TIFAC-SIDBIBrochure.pdf accessed at 4:06 PM IST.
3. National Horticulture Board (2013) NHB Schemes at a glance, *Indian Horticulture Data Base 2013,* pp 269-276.

*Chapter 12*

# INNOVATIVE MODELS IN AGRI ENTREPRENEURSHIP

**Baljeet Singh**

*Doctoral Scholar, GH Patel Post Graduate Institute of Management, V V Nagar, Gujarat*

The first and foremost trait of a breakthrough innovation is that it should address a challenge, a challenge of the masses, communities, and numbers. This is a task accomplished well immersed the user environment, spending time with the people, a sort of field exercise. This is often called empathy **mapping**.

The next important step is to synthesize what has been observed, heard, felt, and seen in the field and define the challenges properly. Framing the right problem is the only foundation to create the right solution. Synthesise and choose a limited number of needs that need to be satisfied. Empathy mapping is essential to bring out the insights developed through synthesis of information in the field. Problem definition helps in explicit expression of the challenge we are trying to solve based on new understanding. This brings us a discrete problem statement.

**Ideation** is best achieved by brainstorming among a team as big as possible. As Paul Linus says, to have a good idea, you must first have lots of ideas. Though more choices may complicate the situation, but in the long run it proves better. While brain storming should be borne in mind that no idea is bad and there should not be any effort to suppress the idea. However, care should be taken not to deviate from the topic. As the discussion moves on few ideas would come on the top, while others get dropped.

Next step is **prototyping**, converting best ideas into an implementable action plan. A user can interact with prototype. He/she should be able to experience it. A prototype can be made on a wall with charts and post it notes on the charts with clear steps and directions or through a role playing activity or as story board.

Action plan needs to be **piloted** or **tested** which reveal implementation challenges. This is more valid in developing countries like India where ideas look good on paper and in offices but implementation gets stuck because of innumerable challenges like infrastructure, politics, illiteracy etc. After doing this pilot study, you get to know about the necessaty modifications your innovation would require? It is always advisable to make suggested changes in your innovation before the launch.

This chapter will introduce you to some of the innovative start- up entrepreneurs in agri-business arena in India.

## ONGANIC-CONNECTING ORGANIC FARMERS DIRECTLY TO CONSUMERS, ELIMINATING M

**Ekta Kothari Jaju**

**Punam Jaju**

### About the Innovation

ONganic is a social enterprise that connects small and marginal organic farmers with urban consumer, as a producer company, grow, process and package the produce which is sold by ONganic through online and offline platforms (Fig. 12.1).ONganic promotes organically grown indigenous variety of non-perishable food products including cereals, pulses, flours, spices, dairy and other health products. Our unique supply chain enables us to keep our price lower than market and offer affordable organic alternatives to our consumers at their doorsteps while enabling farmers to earn higher incomes.

ONganic is a spin-off of Switch ON (*www.SwitchON.org.in*) and is supported by MSME, Yale School of Management and IIM Calcutta. Additionally ONganic was selected for Egiye Bangla, an entrepreneurship TV reality show by Govt of West Bengal, to identify, recognize & reward aspiring and early-stage entrepreneurs having unique innovative start- up ideas across all sectors.

### Inspiration Behind the Innovation

Currently agriculture is plagued by too many steps and does little value addition - the supply chain is unregulated and un-organised. The food supply chain is very complex in rural areas where small farmers are not benefited in the traditional sales of their produces. There is approx. 400% price hike on produce from farmer to the consumer. Distribution and supply chains are sectioned with 5-7 intermediaries, who add little value.

It is very critical to develop models to connect farmers directly with consumers to ensure a steady income for farmers, and ensure that they are self reliant and self sustainable. Value addition to produce such as, improving produce drying and packaging techniques can be profitable to farmers. In-turn this will reduce farmer debts and incidences of farmer suicides in the country, which are on the rise. It also allows for producers and consumers to connect and come to a common platform through mutual trust and respect.

Organic farming is currently driven mainly by large farmers, as 3rd party certification is very expensive. If small farmers can be aggregated to produce and process organic food directly and supply to the consumers through a unified front, the produce can become more affordable, at the same time benefit the farmers directly.

Initially we organized voluntary groups and organic melas to create awareness with consumers in urban pockets, but very soon realized that unless the marketing activities were not done in a financially sustainable manner, it will not provide a lasting solution. Thus ONganic Foods Pvt Ltd was launched to bridge the gap (Fig. 12.1).

## Challenges Addressed or Problem Solved by the Innovation

The major challenge in the agricultural sector in India is that the small and marginal farmers are fragmented. The resources owned by them are also very low. SwitchOn engaged with the farmers and realized that integration of the small farmers into a group would help in their economic growth also extensive chemical farming is expensive and environmentally taxing in the long run.

The organic farming initiative along with the formation of the farmer producer company helped in the value addition to the product (which is required to surpass the grip of the middlemen). Thus SwitchON has initiated a supply chain innovation, which attempts to create a sustainable agricultural framework for the rural marginal farmers in India. Organic farming was introduced not only to benefit the farmers in the long run, but also to save the environmental resources.

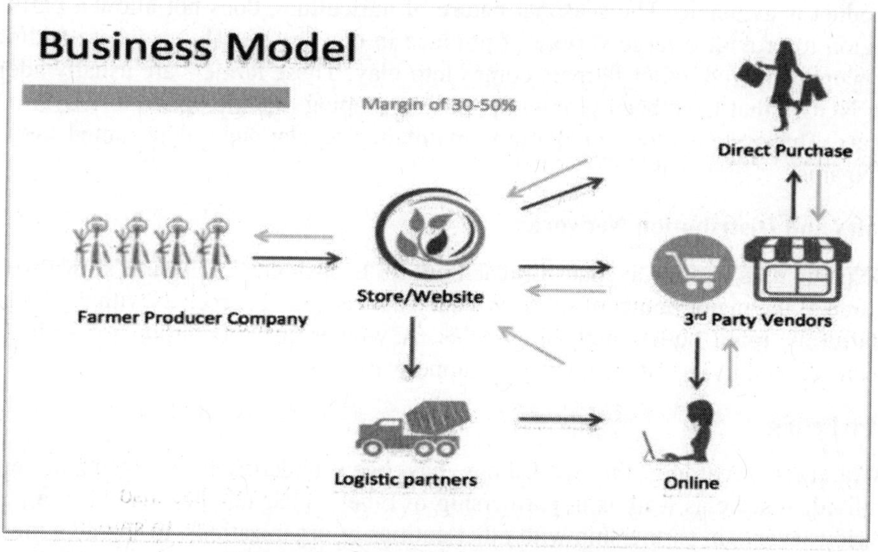


**Figure 12.1:** The ONganic business model

They are also plagued by the interference of the middlemen, who with no value addition sell the product to the consumers at a 400% higher price. The direct linkage of the urban consumers to the farmers would ensure that the maximum benefits are enjoyed by them directly. The traditional methods and modes of marketing is expensive, so it is necessary to look for an economically viable option which will fetch the farmers the best returns for their produces.

The innovative marketing method used by ONganic is inspired from the Tupperware model. Various avenues of marketing channels are used to reach out to the direct consumers. ONganic initiatives address the biggest ache point in organic food chain – reducing costs by connecting the farmers with the consumers and make it available at the consumer's doorstep. Complete end-to-end solution – after understanding customer needs and pain points.

**Farmer Producer Company**

The farmers are integrated into a collective and formed a farmer producer organization. This organization is collectively responsible and is based on community approach model. There are over 100 organic farmers who source their products through is FPO. The food processing like sorting and grading of the products are also done by the women in this community. These farmers have entered into a contract with ONfarm to sell their organically grown produce (under careful due diligence) at a price determined through negotiations involving the farmer himself.

**Supply Chain**

The products (mostly non perishables) are collected and transported to the ONganic Kolkata office. ONfarm track farmers and the season in which a particular type of product is available. The seasonal nature of agriculture, does not allow a particular region to provide a large variety of produce in one season. This is where additional procurement from other farmers comes into play. These farmers are usually adopted by NGO's that have been promoting and developing organic agriculture over a few years. This system allows ONFarm to maintain a regular and uninterrupted supply to ONganic.

**Sales and Distribution Networks**

ONganic uses various avenues to reach out to its customers. ONganic's own website is one of the major points of sales. We have also engaged ourselves with e-commerce platforms, retail outlets and direct sales. A women network group for selling has been adopted in accordance with the Tupperware model.

**Marketing**

Awareness regarding the workshops has been undertaken by ONganic on an individual scale as well as in partnership of others. ONganic has also been a part of various events in partnership with other entities or organizations, to spread a message on a larger and more consolidated front. ONganic has been actively involved in fairs and exhibitions organized in the city, to capitalize on the diverse market that it opens up. We continue to look for ventures like so, for the same purpose.

## Costs, Returns and Profits

Currently ONganic's biggest expense heads are working capital for procurement, marketing, and capital expenditure on community based food processing equipment and storage. ONganic has a margin of approximate 30% on sales, and expecting to break even in 3 years.

## Message to Young Entrepreneurs/Start- Up Agri-Businesses

Agriculture sector provides great opportunities for entrepreneurship, and there is a lot of scope for innovation and value addition. A focus on positively impacting the small farmers can go a long way in addressing one of the biggest challenges we face in our country. Models like Amul, are a great inspiration of entrepreneurship, and something we can all learn from.

## Challenges Faced and Way Out

The traditional modes of marketing are expensive especially in the cities. The organic products are popular among the urban consumers, so to reach out to the customers directly to reap maximum benefits. ONganic was launched to focus on creating a direct market linkage, and eliminates the role of middlemen. The business model identified by ONganic is based on the complete production, processing to marketing process. Thus creation of the whole integrated system was a challenge. The organic farming practices are mostly lost (due to extensive use of modern chemical farming practices) and the traditional knowledge was not enough. The farmer had to be trained in accordance to the modern organic techniques. Extensive research was done on the farming practices and a number of capacity building training programmes were organized. However, the cost of production came down initially (because organic farming uses locally available materials) but the produces required a direct outlet to the customers.

The consumers are aware of the organic products in the market, and they prefer standardized branded products. The fragmented group of farmers individually would not be able to access the end consumers. The solution to this problem was the proper integration of these fragmented farmers in a group, thus farmer producer companies were formed. These organizations are based on community approach.

## In a nutshell

The most interesting thing noticed with the onset of the ONganic venture, the farmer's sense of belonging to their farmland increased. Switch ON has attempted to create a sustainable agriculture ecosystem, which will be beneficial for the farmers in the long run. Farming as an occupation is looked down upon by the new generation in the rural India, with the value addition and organic farming process, it has renewed as a viable rural occupation now. There has been a 15-20% increase in their incomes and the number of livelihood options available also increased. The indigenous methods of farming which were lost and the gene pool which was depleting, SwitchON has helped in the revival. Betterment of the lives of the rural poor and increase in their incomes is major outlook of the venture, thus failure is not an option.

## AGRIBIZ CORPORATION-MULCH LAYING MACHINE

**Praful Patel**

### About the Innovation

The most successful innovation from Agribiz Corporation is mulch laying machine. It is a cost effective tool for getting almost 30% more production in dry-land and semi dry-land areas. Mulch laying machine has been adopted by many farmers not only in Gujarat but across the nation. It is very useful for farmers in water scarce regions.

### Inspiration behind the Innovation

About 20 years ago, plastic mulches made a big entry into the agriculture as a means of moisture conservation, preventing weed growth and maintaining soil temperature but the machine was laid manually which took around. For laying mulch in one acre area 7-10 labor would be required working whole day. If you have to mulch 15 acres, then half of the season is over. After working for six months, we came up with this mulch laying machine, which also makes holes automatically in the polysheet at desired space and of desired size (with an adjustable unit).

### Challenge Addressed by the Innovation

The innovation takes care of acute labor shortage in Indian agriculture in general and for laying mulches in particular.

### Business Model

Production unit is located in Vasad (Anand) Gujarat. Machines are mainly sold through online advertising. Machines are sold in other states also. Personal networks and contacts also bring some customers. Every year around 100 machines are sold.

Each machine is sold for Rs. 65,000. Advisory and consultancy is provided free of cost.

**Cost, Returns and Profits**

The innovator saves around Rs. 20 lakhs a year by selling mulch laying machines.

**Message to Young Entrepreneurs**

Agriculture is a complex domain, but the main problem is labor. Innovations on mechanised farming for small holder's agriculture have immense business opportunities. Any cost effective labor saving technology is bound to meet success in Indian agriculture.

## CROPCONNECT

**Ishira Mehta**

**Punit Jhajaria**

**About the Innovation**

Puneet Jhajharia and Ishira Mehta co-founded CropConnect Enterprises Pvt. Ltd. (CCE) in 2013 to develop demand-driven value chain solutions in agriculture by working closely with farmers and customers. The CCE team has been working in the agriculture value chain sector for the last three years with a focus on using market-led solutions to help connect farmers to consumers, especially urban consumers. This way, the farmer understands what the market trends are and the buyers get the produce they are searching for. CCE strongly believes that such customized solutions help increase market access of the produce, build traceability, decrease wastage and improve farmer incomes. CCE is working towards this goal using two approaches:

1. Developing own agriculture value chains.
2. Providing consulting support to organizations.

**We Build Unique Value Chains**

Over the last two years Puneet and Ishira have travelled over 70,000 kms across 20 states to meet and work with farmers. We saw a trend of farmers shifting to better farming practices (often organic and/or zero budget), growing local and indigenous crops in order to revive their soil and re-instilling pride in farming. But we also found many farmers struggling to sell their products beyond local markets due to a lack of innovative demand-driven business models. We studied close to 10 agricultural value chains ranging from honey to ginger to fruits & vegetable exports and even set up one to gain operational experience an end-to-end value chain for marketing of kiwi-fruit from Arunachal Pradesh farmers to New Delhi markets in 2013.

Our experiences through travels, building networks and assessing numerous value chains, finally culminated into our new venture - Original Indian Table (*www.originalindiantable.com*). It is a unique platform that brings the traditional food and wellness experience for customers across India. We provide curated, traditional and healthy alternatives to common food ingredients and products that are sourced directly from the best farmers of India. We also help customers with nutrient information and recipes on the farm produce.

We sell our products through traditional B2B channels and innovative B2C mechanisms such as the subscription-box model. Recently we launched two boxes called **Himalayan Holiday** and **Southern Harvest** that had hand-picked products sourced from farmer groups in Uttarakhand, Tamil Nadu and Karnataka. The platform is sustained by building strong back-end partnerships with farmer groups/ NGOs for sourcing, and an efficient logistics operation.

We believe this platform will support the revival of the traditional and sustainable food movement in India, help to increase the commercial potential of these crops and provide the growing market segment of health and environmentally conscious consumers in India with domestic options for e.g. domestic millets instead of quinoa imported from Peru.

## Our Consulting Services Support Farmer Groups

CropConnect also provides customized consulting services to funds, international financial institutions, enterprises, government and donors working on sustainable and organic agriculture and with farmer producer organizations (FPOs). The CropConnect team's strength lies in value chain assessments, business plan development and implementation and marketing strategy. CropConnect also develops customized training solutions for FPOs for e.g. it recently developed a training module for FPOs to introduce farmers to the key financial statements and help them develop a business financial model using a case study approach.

CropConnect has worked on projects with The World Bank, KFW, NABARD, StartUp!, Center for Education and Communication (CEC), International Resources for Fairer Trade (IRFT) and International Foundation for Humanistic Development (IFHD).

## Inspiration behind the Innovation

We, Puneet and Ishira, the founders of CCE worked extensively with the Grassroots Business Fund (GBF) and International Finance Corporation (IFC)– the private sector arm of the World Bank Group respectively – before building our own start-up. Through our work experiences, we found it very difficult to find innovative, market driven and investible business models in agriculture. Yet, there were many farmers who were growing sustainable and high quality products. What was missing was that these products were not, often, able to find the right customers. We both also realized that there was a vast disconnect between the people who grow food and the people who buy it in India We felt that people had often forgotten where their food came from and that is why they had so little regard for the people who were growing that food. This is the gap we wanted to fill. So we quit our jobs and started Crop Connect.

India as a country cannot grow or progress if most of its people – who still work in agriculture – cannot find higher incomes and sustainable livelihood. Mass rural to urban migration is an indication of how agriculture is increasingly becoming unviable as a vocation. A strong and productive rural economy is essential for India's sustainable growth and the role of farmers in feeding a growing national population cannot be emphasized more.

Lack of food safety is a new form of urban poverty today with rampant adulteration and extreme pesticide usage. While one part of the solution is to support enlightened farmers in their passion for sustainable agriculture, the other part of the solution is to inform and engage end customers. Breaking the current disconnect that consumers have with their food and supporting farmers is what drives us everyday.

Above all, working with farmers has been a deeply introspective and transformative experience for us. Their deeper understanding of life and ability to live in harmony with the elements feeds into our lives and makes us better individuals

## Challenges Addressed the Innovation

About 63% of the Indian workforce is engaged in agriculture. Yet it contributes to a mere 16% of the GDP and is growing at only 3%. Poor access to knowledge on sustainable farming practices and infrastructure coupled with growing national demand for food has led to over farming and rampant exploitation of soils. This has resulted in a dip in productivity and returns to farmers leading to 44% of current farmers not wanting to continue farming.

While Indian agriculture is in the state it is today due to a host of commonly cited reasons, such as poorly implemented policies, small land holdings, ineffective extension services, lack good farming inputs and exploitative middlemen- one reason which is often overlooked is the lack of market based approaches to agricultural value chains. Many organizations are working on solving supply side issues but not enough are building demand driven models to reduce the disconnect between farmers and buyers. This is the primary challenge we are trying to address through our work at CropConnect. Be it our own value chain initiatives like Original Indian Table or consulting services, our aim is to remove information asymmetries in the agriculture value chain, empower the farmer with the right business skills, market trends and increased sales and educate the customer on where their food is coming from.

We believe the lives of farmers we work through the Original Indian Table platform will improve in multi-fold ways. One, by helping commercialize regional products we will re-kindle the faith of farmers in sustainable farming. For instance, we successfully supplied regional products from an organic Tamil Nadu farmer group for the first time in the Delhi market, for use in contemporary products like chocolate. Two, as we will directly source from farmers we will cut the many layers of middlemen that exist and give a larger proportion of the end-price to the farmer. Three, we will help to reduce the market risk by offering steady demand and prices for farmers' products. Four, we will bring market feedback directly to the farmers and help them improve their product basket through better seed selection, farming practices or processing facilities. Where viable we will invest our resources to strengthen the supply chain.

Over time these activities will enable farmers to increase income and become proud participants of the Indian economy.

## Business Model

Our business model is a farm to fork model where we source products directly from farmer groups, brand (where ever needed) and package and finally market the products to customers. We pay our farmer groups immediately upon sourcing the products. So far we have been using low cost marketing channels such as social media, phone, SMS and other messenging services like Whatsapp. The infographic below gives more details on how our model operates (Fig. 12.2).

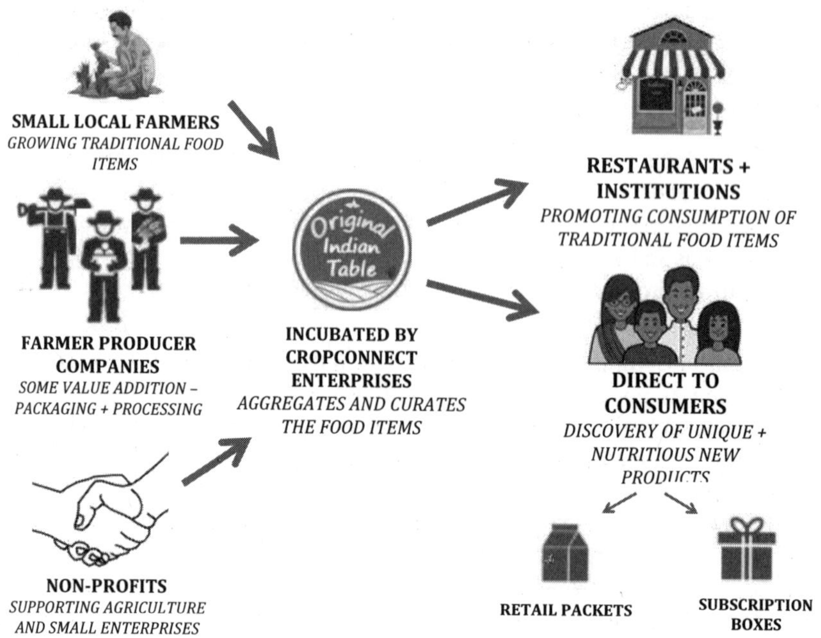


**Figure 12.2:** CropConnect business model

## Costs, Returns and Profits

Having worked in the social enterprise sector in our previous jobs, we have seen many examples of organizations starting off using grants as impetus but without a sustainable business model in mind; eventually not being able to sustain organizational growth. While all social impact models need not be fully commercially sustainable and might need some grants in the start- up and growth phase, we strongly believe that the goal should be to create a financially sustainable model for long term impact.

Keeping this in mind we are developing a business model where social impact and financial returns are directly co-related. As business grows we will generate financial returns through economies of scale and impacting more farmers. We will set benchmarks, for sharing value created by the business, between the enterprise and the farmer. For example if we create value by disrupting the current value chain then some %age of that value will be shared with the farmer. Right now we are operating at operation break-even through our activities.

## Message to Young Entrepreneurs/Start-Up Agri-Businesses

Our advice to young entrepreneurs is to focus on 3 things: passion, value creation and action.

**1. Passion:** As Thomas A. Edison once said Many of life's failures are people who did not realize how close they were to success when they gave up. Everything that can go wrong will go wrong in a start- up and in most cases all your assumptions will be proven wrong. In such circumstances it is very easy to give up and lose heart. Passion and belief in one's idea is the key to sustaining one's energy levels amongst all the chaos. We, at CropConnect, started three years ago with the idea of increasing farmer income through market linkages and better post-harvest management. After three years of experimenting with business ideas and several pilots we have finally found the idea we want to take to market. What kept us going all these years was the belief that there are issues and bottlenecks in the current system, which when unlocked will create value for everyone – the farmer, the customer and, of course, CropConnect.

**2. Value creation:** When working on the business idea it is critical to understand what is that unique value proposition that you creating and for whom. This is usually done by understanding your customer, the customer's unfulfilled needs and creating a product or service to cater to those needs. At CropConnect we spoke to several people that are looking for traditional, ancient and healthy food options; and based on our understanding of customer needs we conceptualized Original India Table.

**3. Action**: Sometimes the best way to understand all of this is to quickly prototype your idea, instead of desktop or field research.

## Challenges Faced and Way Out

Most entrepreneurial ventures, especially in a country like India, face innumerable hurdles – from identifying the idea to focus efforts on to relentless administrative and bureaucratic challenges, from not being able to hire the right people to consistently making do with a shoe-string budget. But agriculture – one of the most difficult sectors – has even greater roadblocks.

### Awareness of Safe and Sustainable Produce

We need to create awareness in the minds of millions of customers who are not used to differentiating between farm produce, to actually demand and pay for safe and sustainable produce.

### Farmer Trust

Winning the trust of farmers is another challenge we have faced as we engage with them. Over the years they have been approached by so many organizations that the initial approach of farmers is one of scepticism – doubting any favourable action from urban companies. Through all the challenges, what has helped us is clarity of purpose and persistence. People and situations start supporting you if you persist with a genuine intention. This has been our experience.

### Case Study

When we started CropConnect, one of our first projects was to try and link farmers in remote Arunachal Pradesh, one of the most geographically challenging states of India, to markets in the national capital, Delhi. CCE worked on a four-month pilot project with kiwi growing farmers to help them bring the fruit to the Delhi market. At that time, we knew nothing about kiwifruits and had never worked in the north east of India. We faced challenges at every step in our project. It took over 24 hours from Delhi to get to the nearest town for our work, we constantly battled bad roads, limited goods transport infrastructure and intermittent telephone availability.

But undaunted by these difficulties, we managed to build relationships with the local farmers and built the entire supply chain from post-harvest processing, branding, packaging, and transport operations. We spent two months at the farm gate of 15 farmers building their capacity for primary value addition such as grading and sorting. We manually processed 25 tons of kiwifruit and successfully marketed 8.5 tons of them in the Delhi market fetching prices equal to imported kiwis. Today the farmers still call us and are selling directly to Delhi

## Other Information

We personally use every product before we sell it. Our vision is to make food personal, and celebrate its diversity. One's association with food must not be that of a mass assembly line. CCE is vested and passionate about its products. We do not sell something where we do not have a connection with the farmer group.

We try to keep the tone and verve of our product as close as possible to the ground and do not over package the product. With our simplicity of approach we hope to bring lower costs for the consumers and higher prices for the farmer.

## NATUR OMA

**Titu Yoko**

### Brief Introduction about Innovation

Natur Oma is into collection of organically produced agri-horti products of rural farmers of North-east especially of Arunachal Pradesh and marketing it raw or as value added products without losing on nutritive value. Under banner of NaturOma we make any many healthy and delicious food items from anything healthy and exotic grains organically produced by rural farmers.

### Inspiration behind the Enterprise

I personally believe one should not work for money alone, there is humanity beyond money. Money is by-product of service one renders. When we started Natur Oma, earning profit was never in our mind. While working with the rural farmers and unemployed youths since last ten years we realized that there is a need of tapping the untapped resources for sustainable development in the region or state particularly. Many well thought or noble intended policies and programme have failed due to one or other reason. There is minimal presence of corporate world in the area. Our farmers need backward and forward support (1) regular market support (2) hassle free financial assistance when needed (3) rural friendly suitable technology (both pre and post harvest).

There is peculiar market system in our state for instance the products which come from thousands of kilometres, not fresh and inorganic (grown using chemicals) sell like hot cakes in local market, but our farmers' fresh and organic produce is compelled to be fed to animals. So many of farmers who were doing chemical free farming migrated to towns in search of better livelihood like selling inorganic or unhealthy products as street vendors.

When informed that organic food is in good demand in foreign markets and it is also the food of royal class people, the local farmer responded that such products need to be wrapped in costly or good looking packet. They purchase raw materials from us in five rupees kg and just wrapped in good looking packet and re-sell us at rate thrice of the amount they purchased from us in form of raw materials. So, we (producers) are undernourished and they (buyers) are over nourished (laughs). We have best fruits like orange, kiwi, pineapple, spices like ginger, turmeric, black pepper and king chilli of which we are the highest producer in India

I personally very much passionate about organic products and thus many of my friends address me as Ms Organic. So ours is an attempt to with three objectives: (1) To encourage farmers doing chemical free farming through buy back assurance of their farm produce (2) To create greater employment opportunities for unemployed youth by engaging them in the process and marketing and (3) To make healthy and organic food affordable to all classes of people. Thus preventing migration of farmers and preserving the environment.

## Challenges Addressed or Problems Solved by the Business

No challenges mean no innovations. All wonderful innovations of the world are come out of challenges. But as I said we haven't done any great innovation yet but keep experimenting for better solutions to the challenges we and our communities face. Challenges are on many fronts including-financial, technology, organizing the unorganized (farmers and youths) and tapping the market – the taste or liking of people, convincing people to adopt new ideas or new concepts.

*Jo dikhta hai woh bikta hai* is the slogan in today's market. People don't bother what the product is made of and how it is going to harm our health in long run as long is good looking and well packaged. When asked to wholesaler, retailer and mall owners, they all said we keep the products what our consumer demands so if you want your products to be sold first convince our consumer. So we started door to door marketing through our market representatives, members of our Self Help group and youth club members. Now after two years of our hard efforts we have been able to win the confidence of consumers and now our products are sold in many outlets small to big.

In addition, our youth members were sent for training on network marketing so that we can reach masses and create more employment opportunities for our people. Social media marketing and e-marketing is also in line. We also send some of our female youths to undergo hospitality training so that they can do catering of our products. But above all branding is always a challenge. Public always loves to buy branded products so we are concentrating on branding our products so that we can tap international markets. So we are trying to develop speciality foods like that for diabetics

## Business Model

Our model is a journey rather than a destination and is ever evolving. One thing ingrained in our DNA is respecting our commitments. We follow certain ethics and integrity policies in our dealings. We compete with ourselves to deliver best service to our valued consumers. Organizing the entire value chain from farm to fork is a really tough task especially in a state like Arunachal Pradesh where there are no basic amenities like *pakka* road, regular transportation facilities, electricity, internet or mobile connectivity. We felt the need of organizing farmers into formal groups in the form of farmer club one village- one farmer club and five to eight villages form one cluster. Each cluster has one facilitation centre (FC) which is a sort of one stop solution of all farmers' needs. We do regular follow up to motivate them and give technical or management support. We have opened saving accounts for each individual farmer. All transactions are routed through their bank accounts. Farmers can also avail the KCC loan if needed based on his bank account records. Farmers will plant what we need so that we can assure them of buy back. We collect their produce in our FC then transport it to processing centre to make finish products.

### Message to Young Entrepreneurs/Start-Ups

The message we would like to convey to our young entrepreneurs is that start with an open minded and successful entrepreneurs need not have lots of funds in saving accounts. To me entrepreneur means anyone who brings significant impact or positive changes to the society and mankind. Secondly, your unhappy customers are your best guides. Try to see an opportunity in every challenge. Problems come with solutions, just see and think out of the box and act confidently. If you know where you are going, whole world will stand aside to let you go. If your intention is genuine and work is noble there is no dearth of good people who will come for your help.

I have numbers of like-minded people as mentors, friends within or outside the state who keep encouraging, motivating, guiding us and extend all support whenever we need like Rajeev Takuk Galling, CEO of Livelihood Mission (Arunachal Pradesh), Aditya Tyagi, Ashish Kanwal, Anand Bernawal, Dheeraj Sharma and of course a man behind who is letting you know about us through the book which you are reading right now (Baljeet) there are many more numbers whom I could not mention but they in my heart and mind.

### Challenges Faced How these were Tackled

Challenges and problems are our best friends which either breaks us or we break them. When we break them we are called as successful and when they (problems) break us we are named as failure. Temporary failure teaches us more than what we learnt from success. When we face any challenges we look for solutions to overcome them which make our brains and bodies active and active brains hardly get sick. Every day is challenging which raises our creativity. We keep looking for alternative solutions which are better suited for our farmers and youths. For example our government installed cold storage for storing of farmer's produce which cost lakhs of rupees but almost all cold stores are now defunct as there is no electricity and it takes two days to reach the cold store by foot march. We find the solution in the form of Zero Energy Cool Chamber (ZECC) which can be locally made with minimal investment (Rs. 12,000) and doesn't require electricity.

■■■■